T0306022

Mechanical Behavior
at Small Scales —
Experiments and Modeling

MATERIALS RESEARCH SOCIETY
SYMPOSIUM PROCEEDINGS VOLUME 1224

Mechanical Behavior at Small Scales — Experiments and Modeling

Symposium held November 30 – December 3, Boston, Massachusetts, U.S.A.

EDITORS (Symposium FF):

Jun Lou
Rice University
Houston, Texas, U.S.A.

Erica Lilleodden
GKSS Forschungszentrum
Geesthacht, Germany

Brad L. Boyce
Sandia National Laboratories
Albuquerque, New Mexico, U.S.A.

Lei Lu
Chinese Academy of Sciences
Shenyang, China

EDITORS (Symposium GG):

Peter M. Derlet
Paul Scherrer Institut
Villigen, Switzerland

Daniel Weygand
Karlsruhe Institute of Technology
Karlsruhe, Germany

Ju Li
University of Pennsylvania
Philadelphia, Pennsylvania, U.S.A.

Michael Uchic
Air Force Research Laboratory
Dayton, Ohio, U.S.A.

Eric Le Bourhis
Université de Poitiers
Poitiers, France

Materials Research Society
Warrendale, Pennsylvania

CAMBRIDGE
UNIVERSITY PRESS

University Printing House, Cambridge CB2 8BS, United Kingdom

One Liberty Plaza, 20th Floor, New York, NY 10006, USA

477 Williamstown Road, Port Melbourne, VIC 3207, Australia

314-321, 3rd Floor, Plot 3, Splendor Forum, Jasola District Centre, New Delhi - 110025, India

79 Anson Road, #06-04/06, Singapore 079906

Cambridge University Press is part of the University of Cambridge.

It furthers the University's mission by disseminating knowledge in the pursuit of education, learning and research at the highest international levels of excellence.

www.cambridge.org
Information on this title: www.cambridge.org/9781605111971

Materials Research Society
506 Keystone Drive, Warrendale, PA 15086
http://www.mrs.org

© Materials Research Society 2010

First published 2010
First paperback edition 2012

Single article reprints from this publication are available through University Microfilms Inc., 300 North Zeeb Road, Ann Arbor, MI 48106

CODEN: MRSPDH

A catalogue record for this publication is available from the British Library

ISBN 978-1-605-11197-1 Hardback
ISBN 978-1-107-40803-6 Paperback

CONTENTS

* Invited paper

v

* Invited paper

PREFACE

Symposia FF: "Mechanical Behavior of Nanomaterials—Experiments and Modeling" and GG: "Plasticity in Confined Volumes— Modeling and Experiments," were presented Nov. 30–Dec. 4, 2009 at the 2009 MRS Fall Meeting in Boston, Massachusetts. Symposium GG began with a very attractive tutorial on experimental and simulation methods for the study of plasticity in small volumes. The strongly related themes of these two symposia motivated the publication of a single proceedings volume. These two symposia each brought experimentalists and modelers together within a single forum to exchange their ideas about the mechanical behaviour of materials where size, be it microstructural or geometric, plays an important role.

Symposium FF was focused on the understanding of the mechanical behavior of nanostructured materials, such as nanoscale thin films, nanowires, nanotubes, and nanoparticles, as well as nanoporous, nanograined and nanotwinned materials. Such materials with sub-micron length scales are important building blocks for next-generation functional devices and materials systems. In order to help them fulfill their promise, mechanics at small-length scale must be carefully investigated to understand the deformation and failure mechanisms of these material entities.

Symposium GG was focused on understanding how micron and sub-micron external and internal micro-structural length scales control the mechanical behavior, such as strength and ductility, of materials. Modern atomistic and mesoscopic simulation methods have elucidated a diverse range of atomic and meso-scale processes that can contribute to the emergent plasticity of such complex materials, ranging from dislocation dynamics in micron-sized confined volumes, and the interaction between dislocations and grain boundaries in bulk and thin-film nanocrystals, to atomic scale activity associated with grain boundary accommodation processes, as well as shear-transformation zones in metallic glasses. With the development of miniaturized mechanical testing facilities, as well as leading edge ex-situ and in-situ methods, it now becomes experimentally possible to directly probe both the spatial and temporal dynamics of such processes.

The papers herein span a representative range of topics, which have been organized in four general topic areas: nanostructured materials, polymers & composites, simulations & modelling, and microcompression & nanoindentation.

We would like to thank the authors for their contributions to this volume, as well as the presenters and session chairs for their involvement in the symposia. We greatly appreciate the excellent administrative and technical support provided by the MRS staff, which was a major contribution to the success of these symposia. We also want to extend special thanks to the Institute of Metal Research, Chinese Academy of Sciences and Nanofactory Instruments, Inc. for their financial support of symposium FF.

Jun Lou
Erica Lilleodden
Brad L. Boyce
Lei Lu
Peter M. Derlet
Daniel Weygand
Ju Li
Mike D. Uchic
Eric Le Bourhis

July 2010

MATERIALS RESEARCH SOCIETY SYMPOSIUM PROCEEDINGS

MATERIALS RESEARCH SOCIETY SYMPOSIUM PROCEEDINGS

Volume 1218E —Energy Harvesting—From Fundamentals to Devices, H. Radousky, J.D. Holbery, L.H. Lewis, F. Schmidt, 2010, ISBN 978-1-60511-191-9
Volume 1219E —Renewable Biomaterials and Bioenergy—Current Developments and Challenges, S. Erhan, S. Isobe, M. Misra, L. Liu, 2010, ISBN 978-1-60511-192-6
Volume 1220E —Green Chemistry in Research and Development of Advanced Materials, W.W. Yu, H. VanBenschoten, Y.A. Wang, 2010, ISBN 978-1-60511-193-3
Volume 1221E —Phonon Engineering for Enhanced Materials Solutions—Theory and Applications, 2010, ISBN 978-1-60511-194-0
Volume 1222— Microelectromechanical Systems—Materials and Devices III, J. Bagdahn, N. Sheppard, K. Turner, S. Vengallatore, 2010, ISBN 978-1-60511-195-7
Volume 1223E —Metamaterials—From Modeling and Fabrication to Application, N. Engheta, J. L.-W. Li, R. Pachter, M. Tanielian, 2010, ISBN 978-1-60511-196-4
Volume 1224— Mechanical Behavior at Small Scales—Experiments and Modeling, J. Lou, E. Lilleodden, B. Boyce, L. Lu, P.M. Derlet, D. Weygand, J. Li, M.D. Uchic, E. Le Bourhis, 2010, ISBN 978-1-60511-197-1
Volume 1225E —Multiscale Polycrystal Mechanics of Complex Microstructures, D. Raabe, R. Radovitzky, S.R. Kalidindi, M. Geers, 2010, ISBN 978-1-60511-198-8
Volume 1226E —Mechanochemistry in Materials Science, M. Scherge, S.L. Craig, N. Sottos, 2010, ISBN 978-1-60511-199-5
Volume 1227E —Multiscale Dynamics in Confining Systems, P. Levitz, R. Metzler, D. Reichman, 2010, ISBN 978-1-60511-200-8
Volume 1228E —Nanoscale Pattern Formation, E. Chason, R. Cuerno, J. Gray, K.-H. Heinig, 2010, ISBN 978-1-60511-201-5
Volume 1229E —Multiphysics Modeling in Materials Design, M. Asta, A. Umantsev, J. Neugebauer, 2010, ISBN 978-1-60511-202-2
Volume 1230E —Ultrafast Processes in Materials Science, A.M. Lindenberg, D. Reis, P. Fuoss, T. Tschentscher, B. Siwick, 2010, ISBN 978-1-60511-203-9
Volume 1231E —Advanced Microscopy and Spectroscopy Techniques for Imaging Materials with High Spatial Resolution, M. Rühle, L. Allard, J. Etheridge, D. Seidman, 2010, ISBN 978-1-60511-204-6
Volume 1232E —Dynamic Scanning Probes—Imaging, Characterization and Manipulation, R. Pérez, S. Jarvis, S. Morita, U.D. Schwarz, 2010, ISBN 978-1-60511-205-3
Volume 1233— Materials Education, M.M. Patterson, E.D. Marshall, C.G. Wade, J.A. Nucci, D.J. Dunham, 2010, ISBN 978-1-60511-206-0
Volume 1234E —Responsive Gels and Biopolymer Assemblies, F. Horkay, N. Langrana, W. Richtering, 2010, ISBN 978-1-60511-207-7
Volume 1235E —Engineering Biomaterials for Regenerative Medicine, S. Bhatia, S. Bryant, J.A. Burdick, J.M. Karp, K. Walline, 2010, ISBN 978-1-60511-208-4
Volume 1236E —Biosurfaces and Biointerfaces, J.A. Garrido, E. Johnston, C. Werner, T. Boland, 2010, ISBN 978-1-60511-209-1
Volume 1237E —Nanobiotechnology and Nanobiophotonics—Opportunities and Challenges, 2010, ISBN 978-1-60511-210-7
Volume 1238E —Molecular Biomimetics and Materials Design, J. Harding, J. Evans, J. Elliott, R. Latour, 2010, ISBN 978-1-60511-211-4
Volume 1239— Micro- and Nanoscale Processing of Biomaterials, R. Narayan, S. Jayasinghe, S. Jin, W. Mullins, D. Shi, 2010, ISBN 978-1-60511-212-1
Volume 1240E —Polymer Nanofibers—Fundamental Studies and Emerging Applications, 2010, ISBN 978-1-60511-213-8
Volume 1241E —Biological Imaging and Sensing using Nanoparticle Assemblies, A. Alexandrou, J. Cheon, H. Mattoussi, V. Rotello, 2010, ISBN 978-1-60511-214-5
Volume 1242— Materials Characterization, R. Pérez Campos, A. Contreras Cuevas, R.A. Esparza Muñoz, 2010, ISBN 978-1-60511-219-0
Volume 1243— Advanced Structural Materials, H.A. Calderon, A. Salinas-Rodríguez, H. Balmori-Ramirez, J.G. Cabañas-Moreno, K. Ishizaki, 2010, ISBN 978-1-60511-220-6

Prior Materials Research Society Symposium Proceedings available by contacting Materials Research Society

Nanostructured Materials

Mater. Res. Soc. Symp. Proc. Vol. 1224 © 2010 Materials Research Society   1224-FF10-37

Variable Elastic-Plastic Properties of the Grain Boundaries and Their Effect on the Macroscopic Flow Stress of Nanocrystalline Metals

Malgorzata Lewandowska , Romuald Dobosz , Krzysztof J. Kurzydlowski

Warsaw University of Technology, Faculty of Materials Science and Engineering,

Woloska 141, 02-507 Warsaw, POLAND.

ABSTRACT

The paper reports new experimental results describing properties and microstructure of nanocrystalline metals. Nano- and sub-micron aluminium has been produced by hydrostatic extrusion at ambient tempearture. The structures have been quantified in terms of size of grains and misorientation of the grain boundaries. Different average size of grains, variable normalized width of grain size distribution and changing grain boundary misorientation distribution functions have been revealed depending on processing parameters. The results of the tensile tests showed that the average grain size, grain size distribution and the distribution function of misorientation angles influence the flow stress of obtained nano-metals.

In order to explain the observed difference in the properties of nano- and micro-sized aluminium alloys, a Finite Element Method models have been developed, which assumes that both grain boundaries and grain interiors may accommodated elastic and non-linear plastic deformation. These models assumed true geometry of grains (which differed in size and shape). Also, variable mechanical properties of grain boundaries have been taken into account (elastic modulus, yield strength and work hardening rate). The results of modelling explain in a semi-quantitative way macroscopic deformation of nano-crystalline aggregates. In particular, they illustrate the importance of the interplay between properties of grain boundaries and grain interiors in elastic and plastic regime.

INTRODUCTION

A vast majority of engineering materials are used in polycrystalline form which makes grain boundaries one of the most important microstructural elements. Their role is particularly important in the case of fine grained and nanocrystalline materials as their surface area per unit volume is substantially greater than in conventional micro-structured polycrystalline counterparts. This can be supported by simple geometric considerations which indicate that for an average grain size of 10 nm the volume fraction of atoms located at grain boundaries is 25% assuming 1 nm thickness of grain boundaries. As a result, the properties of nanocrystalline materials are largely governed by grain boundaries.

The effect of the grain boundaries on flow stress of metals has been a subject of numerous experimental and theoretical studies [1-2]. It is usually described by the Hall-Petch relationship which predicts linear dependence of the flow stress on the inverse square root of grain size. Hall-Petch relationship was experimentally proven for a wide range of materials with average grain sizes ranging from several hundreds of microns to dozens of nanometres.

However, it has been found, in this context, that below a certain critical value of grain size (typically 10-30 nm) the inverse Hall-Petch relationship is observed [3].

Recently, remarkable progress in the understanding of the effect of grain boundaries has been made due to advances in modelling of plastic behaviours of nano-polycrystalline aggregates. In particular, so called composite model was successfully used to explain inverse Hall-Petch relationship [4]. In this approach, nanocrystalline material is considered as a composite consisting of non-zero-thickness grain boundaries and grain interiors, which have different mechanical properties [5]. Despite these achievements, some key questions with regard to the role of grain boundaries remain unanswered and quantification of their effect on the macroscopic properties is still a challenge.

EXPERIMENTAL

In order to experimentally evaluate the effect of grain boundaries on the macroscopic flow stress of aluminium alloys, various sub-micro grained structures were obtained in technically pure 1050 aluminium via processing by hydrostatic extrusion and subsequent low-temperature annealing at temperature ranging from 100 to 300°C. The samples were hydrostatically extruded to 3 mm wire in multistep extrusion processes starting with either 50 or 20 mm billets (a total true strain of 5.4 or 3.8, respectively). They were water cooled at the exit of the die to minimize the effect of adiabatic heating during the process. More information on the processing procedure can be found elsewhere [6].

The obtained microstructures were evaluated qualitatively by TEM observations and quantitatively using computer aided image analysis. Specimens for TEM studies were cut perpendicularly to the extrusion axis. The foils have been examined in a Jeol JEM-1200 electron microscope operated at 120 kV. The grain size was described using the equivalent grain diameter d (defined as the diameter of a circle of equal area to the surface area of a given grain) and variation coefficient CV(d) defined as a ratio of standard deviation SD(d) to the mean value. The misorientation angles were determined for a population of randomly selected grain boundaries. Crystallographic orientations of individual grains were calculated from Kikuchi lines patterns obtained in TEM by convergent beam diffraction. These patterns were subsequently used to calculate misorientations across the boundaries. For each annealing temperature, a population of at least 100 grain boundaries was analyzed.

As HE processed billets are relatively large in dimensions, it was possible to characterize the mechanical properties of the processed materials in tensile tests with relatively good statistics. Tensile tests were conducted at room temperature and at initial strain rate of 10^{-3} s^{-1}.

Finite Element Method, FEM, models, which have been developed in the present study, assume that both grain boundaries and grain interiors may accommodated elastic and non-linear plastic deformation. These models are based on true geometry of grains extracted from TEM images. The grains differed in size, size diversity and shape. Their populations in a given model were described by the average grain diameter, E(d) and coefficient of variation, CV(d). Also, variable mechanical properties of grain boundaries have been taken into account (i.e. elastic modulus, yield strength and work hardening rate). For FEM calculations, mechanical properties of grain interiors were kept constant whereas mechanical properties of grain boundaries were systematically varied to evaluate their effect on the macroscopic reaction of the polycrystalline aggregates to the applied force in a tensile test.

4

RESULTS

Hydrostatic extrusion results in a significant grain refinement in metals as was already confirmed for a number of alloys, e.g. aluminium [7,8], titanium [9], steels [10,11]. In the case of technically pure aluminium, the HE induced microstructures consist of fairly equiaxial, well developed grains almost free of dislocations, as illustrated in figure 1. The grains are slightly elongated in the direction parallel to extrusion direction (figure 1c). However, it should be noted that depending on processing parameters (applied strain and annealing conditions), the microstructures differ in the size of grains which usually is described using an average value.

Figure 1. Typical TEM micrographs of 1050 aluminium processed by hydrostatic extrusion with a true strain of 2.7(a) and 3.8 (b,c). (a,b) transverse section, (c) longitudinal section.

In this context, it should be noted that the use of an average value of grain size is fully justified only under assumption that the size of grains can be described by the same size distribution function. On the other hand, polycrystalline materials must be viewed as stochastic populations of grains which may differ significantly in size and shape. As a result, the same average value may ascribe for far different grain structures, as illustrated in figure 1. In addition, individual grain boundaries may have different properties due to different misorientation angles. The results of measurements of various microstructural parameters (namely average grain size, grain size coefficient of variation and distribution of misorientation angles) for the samples investigated in this study are listed in Table I.

Table I. Microstructural parameters of 1050 aluminium after various processing conditions

spl	processing parameters	YS [MPa]	ε_t [%]	grain size		fraction of grain boundaries [%]		
				E(d) [nm]	CV(d)	$\theta<5°$	$5°<\theta<15°$	$\theta>15°$
1	as-received	149	7.1	920	0.41	-	-	-
2	HE, $\varepsilon=1.4$	176	6.3	610	0.43	65	25	10
3	HE, $\varepsilon=2.7$	175	8.1	580	0.42	22	33	45
4	HE, $\varepsilon=3.8$	202	6.5	600	0.36	20	20	60
5	HE, $\varepsilon=5.4$	205	2.8	320	0.40	8	20	72
6	HE, $\varepsilon=5.4$ + 200°C for 1 h	167	3.3	450	0.50	7	15	77
7	HE, $\varepsilon=5.4$ + 300°C for 3 hrs	53	26.9	35,000	0.45	-	-	-

It should be further noted that grain structures produced by hydrostatic extrusion, alone or in combination with annealing, differ significantly not only in grain size but also in grain size diversity (the variation coefficients of grain size vary from 0.36 to 0.50) and the distribution of misorientation angles of the grain boundaries.

The experimental values of yield strength for sub-micro grained 1050 aluminium plotted in the coordinates of Hall-Petch equation (against inverse square root of the average grain size), show a significant scattering from the linear dependence, as presented in figure 2. For the same average grain size, the samples exhibit far different properties or the other way round the same yield strength is observed for far different grain sizes. The in-depth analysis of microstructural parameters, shown in Table I, indicates that grain size diversity expressed by variation coefficient significantly influences the flow stress, as illustrated in figure 2. The Hall-Petch line is drawn for the data points obtained for samples with CV~0.41. It can be noted that the flow stress for CV~0.50 diverts from the Hall-Petch line downwards. On the other hand, the flow stress for CV=0.36 is shifted upwards. The character of grain boundaries plays also a significant role.

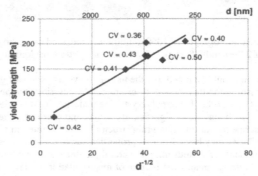

Figure 2. Plot of Hall-Petch relationship for 1050 aluminium processed by hydrostatic extrusion and subsequent annealing

In order to have a better insight into the role of grain boundaries on flow stress of nanocrystalline materials, a comprehensive FEM model (figure 3a) has been developed which takes into account such features of nano-polycrystalline metals as: (1) variable geometry of grains; (2) varied mechanical properties (elastic modulus, yield stress and work hardening rate) of grain boundaries and (3) grain size dependant flow stress of grain interiors. The detail model description can be found elsewhere [12].

FEM computations for different combination of the properties of grain boundaries have been performed to obtain macroscopic stress-strain curves (the flow stress of nano-polycrystalline aggregates was normalized to the flow stress of micro-grained counterpart), as exemplified by the ones presented in figure 3b. The computed changes in the macroscopic Young modulus and flow stress of nano-polycrystalline aggregate as a function of varying grain boundary Young modulus (E^B) normalized to the Young modulus of the grain interior (E^I) are shown in figure 4. It can be noted that the effect of grain boundary Young modulus is particularly pronounced when grain boundaries are less stiff than grain interiors. In such a case, a substantial decrease in Young modulus and macroscopic flow stress of nano-polycrystalline aggregate is observed.

Figure 3. FEM model of nano-polycrystalline metals (a) and macroscopic stress-strain curves of

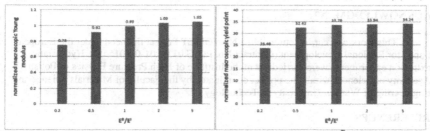

Figure 4. The effect of varying value of grain boundary Young modulus, E^B, normalized to the Young modulus of the grain interior, E^I on (a) Young modulus and (b) yield point of nano-polycrystalline aggregate
Young modulus of the grain interior, E^* on (a) Young modulus and (b) yield point of nano-polycrystalline aggregate

Figure 5 illustrates the effect of grain boundary yield strength R_e^G (the same in normal and in-plane directions) on macroscopic Young modulus, strain hardening rate and yield strength of nano-polycrystalline aggregate. A profound reduction in macroscopic Young modulus and yield strength is observed for low values of grain boundary strength.

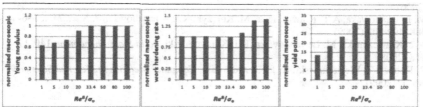

Figure 5. The effect of varying grain boundary flow stress, R_e^B, normalized to the flow stress of the monocrystal of analyzed material, σ_0 on: (a) Young modulus, (b) work hardening rate and (c) yield point of nano-polycrystalline aggregate

The above described findings have implications to understanding the mechanics of nano-metals. It is shown that low macroscopic Young modulus and yield strength in nano-metals might be heavily dependent on the mechanical properties of grain boundaries. This is due to the topology of grain boundaries which in a polycrystalline aggregate form a percolated structure.

CONCLUSIONS

It is shown here that macroscopic tensile behaviours of fine grained polycrystals depend on the average grain size, grain size distribution and the distribution function of misorientation angles. Experimental and numerical modeling results also reveal a significant impact of grain boundary properties on macroscopic deformation of nano-crystalline aggregates. In particular, they illustrate the importance of the interplay between properties of grain boundaries and grain interiors in elastic and plastic regime.

ACKNOWLEDGMENTS

This work was carried out within a NANOMET Project financed under the European Funds for Regional Development (Contract No. POIG.01.03.01-00-015/08). Hydrostatic extrusion experiments were carried out at the Institute of High Pressure Physics of Polish Academy of Sciences within the project coordinated by the Faculty of Materials Science and Engineering of Warsaw University of Technology.

REFERENCES

1. N. Hansen, Mat. Sci. Eng. A409, 39-45 (2005)
2. K.J. Kurzydlowski, B. Ralph, J.J. Bucki, A. Garbacz, *Mater. Sci. Eng.* **A205**, 127-132 (1996)
3. J. Schiotz, F.D. di Tolla, K.W. Jacobsen, *Nature* **391**, 561-563 (1998)
4. R. A. Masumura, P. M. Hazzledine and C. S. Pande, *Acta mater.* **46**, 4527-4534 (1998)
5. X. S. Kim, Y. Estrin, *Acta mater.* **53**, 765-772 (2004)
6. M. Lewandowska, K.J. Kurzydlowski, *J. Mat. Sci.* **43**, 7299-7306 (2008)
7. K. Wawer, M. Lewandowska, K.J. Kurzydlowski, *Mat. Sci. Forum* **584-586**, 541-54 (2008)
8. M. Lewandowska, W Pachla, K.J. Kurzydlowski, *Int. J. Mat. Res. (formely Z.Metallkd)* **98**, 172-177 (2007)
9. H. Garbacz, M. Lewandowska, W. Pachla, K.J. Kurzydlowski, *J Microscopy* **223**, 272-274 (2006)
10. M. Lewandowska, A.T. Krawczynska, M. Kulczyk, K.J. Kurzydlowski, *J. Nucl. Mat.* **386-388**, 499-502 (2009)
11. A.T. Krawczynska, M. Lewandowska, K.J. Kurzydlowski, *Solid State Phenomena* **140**, 173-178 (2008)
12. R. Dobosz, T. Wejrzanowski, K. J. Kurzydlowski, *Computer Methods in Material Science* **9**, No. 1, (2009)

Mater. Res. Soc. Symp. Proc. Vol. 1224 © 2010 Materials Research Society   1224-FF10-32

Deformation behavior of nanocrystalline Co-Cu alloys

Motohiro Yuasa[1], Hiromi Nakano[2], Kota Kajikawa[1], Takumi Nakazawa[1], Mamoru Mabuchi[1]

[1] Graduate School of Energy Science, Kyoto University, Yoshidahonmachi, Sakyo-ku, Kyoto, 606-8501, Japan
[2] Cooperative Research Facility Center, Toyohashi University of Technology, 1-1 Hibarigaoka, Tempaku-cho, Toyohashi 440-8580, Japan

ABSTRACT

Three kinds of nanocrystalline Co-Cu alloys: a nanocrystalline Co-Cu alloy with nanoscale lamellar structure, a supersaturated solid solution Co-Cu alloy and a nanocrystalline two-phase Co-Cu alloy were processed by electrodeposition, and their mechanical properties were investigated at room temperature. These nanocrystalline Co-Cu alloys showed the high hardness and the low activation volume. The mechanical properties of the nanocrystalline Co-Cu alloys strongly depended on the grain boundary characteristics. Molecular dynamics simulations were performed in the two-phase nanocrystalline Co-Cu alloy to investigate the dislocation emission at the Co/Cu interface. The MD simulations showed that the stacking faults, which are generated by the intense geometrical strain at the Co/Cu interface, play an important role in the dislocation emission.

INTRODUCTION

Co alloys are one of promising metallic materials because they exhibit high heat resistance, ferromagnetism and so on. For various applications, it is desirable to improve the mechanical properties of Co alloys. Nanocrystallization can give rise to a significant enhancement of mechanical properties in metallic materials. However, nanocrystalline metals tend to be very brittle with a ductility of less than a few percent in tensile tests [1,2], due to the absence of dislocation activity [3]. It is accepted that nanocrystalline metals show the high hardness (high strength) and the low activation volume [4-9]. These features of nanocrystalline metals are attributed to emission of dislocations at the grain boundaries, and the grain boundaries play a critical role in deformation of nanocrystalline metals. Hence, it is required to develop nanocrystalline Co alloys with unique grain boundaries for enhancement of the mechanical properties.

In the present work, three kinds of nanocrystalline Co-Cu alloys are processed by electrodeposition, and their mechanical properties are investigated at room temperature. In addition, molecular dynamics (MD) simulations are performed in the nanocrystalline two-phase Co-Cu alloy to investigate the dislocation emission at the Co/Cu interface.

EXPERIMENTAL

Three kinds of nanocrystalline Co-Cu alloys, that is, a nanocrystalline Co-Cu alloy with nanoscale lamellar structure, a supersaturated solid solution Co-Cu alloy and a nanocrystalline two-phase Co-Cu alloy were processed by electrodeposition [10]. The electrolyte composition was $CoSO_4 \cdot 7H_2O$ (1 M) and $CuSO_4 \cdot 5H_2O$ (0.025 M). Microstructure of the Co-Cu alloys was investigated by transmission electron microscopy. Mechanical properties of the Co-Cu alloy were investigated by the hardness and tensile tests at room temperature. The hardness tests were performed with a diamond Berkovich tip at constant loading rates of 13.24, 1.324 and 0.378 mN/s.

RESULTS AND DISCUSSION

Nanolamellar Co-Cu alloy

A transmission electron microscopy image of a nanocrystalline Co-Cu alloy with nanoscale lamellar structure is shown in Fig. 1. The grain size of the Co-Cu alloy was 110 nm. Note that most of the grains contained a high-density fine nanoscale lamellar structure. In previous studies [11,12], the nanocrystalline Cu with nanoscale twins with a spacing of tens of nanometers was fabricated by electrodeposition. On the other hand, the Co-Cu alloy developed in the present work contained nanoscale lamellar structure with a much smaller spacing of 3 nm.

Figure 1. A transmission electron microscopy image of the nanocrystalline Co-Cu alloy with nanoscale lamellar structure.

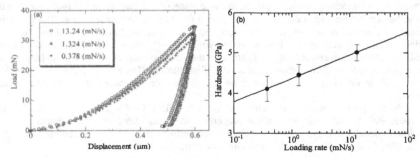

Figure 2. The results of hardness tests for the nanocrystalline Co-Cu alloy with nanoscale lamellar structure, (a) load-displacement curves at three different loading rates and (b) variation in hardness as a function of loading rate.

Load-displacement curves obtained from the hardness tests at the three loading rates are shown in Fig. 2(a). The hardness of the nanolamellar Co-Cu alloy was 4.12-5.02 GPa. As shown in Fig. 2(a), a higher load was required at a higher loading rate to impose the same displacement. The variation in hardness as a function of loading rate is shown in Fig. 2 (b). From the results in Fig. 2, the strain rate sensitivity and activation volume were 0.055 and $3.3b^3$ for the nanolamellar Co-Cu alloy, respectively. The activation volume for the nanolamellar Co-Cu alloy is much lower than those for the nanotwin Cu [13]. Clearly, the low activation volume for the nanolamellar Co-Cu alloy is attributed to the nanoscale lamellar structure of 3 nm.

The tensile test showed that the yield (0.2% proof) stress and ultimate tensile strength of the nanolamellar Co-Cu alloy were 1420 and 1875 MPa, respectively. The yield strength is higher than that of nanocrystalline Co with a grain size of 12 nm (= 1002 MPa) [14]. Also, the nanolamellar Co-Cu alloy showed an elongation to fracture of 3.3%, which is larger than those for nanocrystalline metals with a grain size of less than 10 nm and containing no nanotwins [1,2]. It has been demonstrated in nanocrystalline Cu with nanotwins that twin boundaries can act as dislocation sources [12]. Twin boundaries not only behave as obstacles to dislocation motion, but they also serve as dislocation sources during further deformation. Therefore, the high ductility for the nanolamellar Co-Cu alloy is likely to be attributed to boundaries of the nanoscale lamellar structures acting as nucleation/accumulation sites of dislocations as well as the twin boundaries.

Nanocrystalline supersaturated solid solution Co-Cu alloy

A transmission electron microscopy image of a nanocrystalline supersaturated solid solution Co-Cu alloy is shown in Fig. 3. The grain size of the supersaturated solid solution Co-Cu alloy was 41 nm. From the X-ray diffraction spectrum and EDX measurements, the Co-Cu alloy was a supersaturated solid solution by solution of Cu atoms to Co, and the concentration of Cu was 20 at.%.

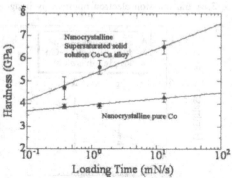

Figure 3. A transmission electron microscopy image of the nanocrystalline supersaturated solid solution Co-Cu alloy

Figure 4. The variation in hardness as a function of loading rate for the nanocrystalline supersaturated solid solution Co-Cu alloy and the nanocrystalline pure Co.

The variation in hardness as a function of loading rate is shown in Fig. 4 for the supersaturated solid solution Co-Cu alloy. For reference, the data for the nanocrystalline pure Co is superimposed in Fig. 4. From the results, the activation volume was $2.5b^3$ for the nanocrystalline supersaturated solid solution Co-Cu alloy and $13.1b^3$ for the nanocrystalline pure Co, respectively. The activation volume for the supersaturated solid solution Co-Cu alloy is much lower than those for nanocrystalline pure Co. Therefore, it is suggested that the solute Cu atoms located at the grain boundaries affect the emission of dislocations at the grain boundaries.

Nanocrystalline two-phase Co-Cu alloy

A transmission electron microscopy image of a nanocrystalline two-phase Co-Cu alloy is shown in Fig. 5. From the X-ray diffraction spectrum and EDX measurements, the Co-Cu alloy consisted of two phases with Co grains and Cu grains. The average concentration of Cu was 38 wt.%. The grain sizes were 141 nm for Co phase and 181 nm for Cu phase, respectively.

Figure 5. A transmission electron microscopy image of the nanocrystalline two-phase Co-Cu alloy.

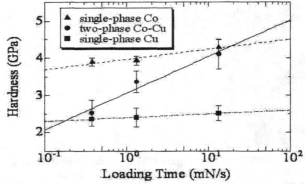

Figure 6. The variation in hardness as a function of loading rate for the nanocrystalline two-phase Co-Cu alloy, the nanocrystalline single phase Co and the nanocrystalline single phase Cu.

The variation in hardness as a function of loading rate is shown in Fig. 6 for the nanocrystalline two-phase Co-Cu alloy, the nanocrystalline single phase Co and the nanocrystalline single phase Cu. From the results, the activation volume was $3.1b^3$ for the nanocrystalline two-phase Co-Cu alloy, $13.1b^3$ for the nanocrystalline single phase Co and $32.8b^3$ for the nanocrystalline single phase Cu, respectively.

It is accepted that nanocrystalline metals exhibit the low activation volume, resulting from the emission of dislocations at the grain boundaries. In the present work, the nanocrystalline Co-Cu alloys with the characteristic grain boundaries: the nanocrystalline Co-Cu alloy with nanoscale lamellar structure, the nanocrystalline supersaturated solid solution Co-Cu alloy and the nanocrystalline two-phase Co-Cu alloy showed much lower activation volumes of about $3b^3$, compared with the nanocrystalline pure metals. The much lower activation volumes of the nanocrystalline Co-Cu alloys with the characteristic grain boundaries cannot be explained only by the refinement of grains. Therefore, it is suggested that the unique grain boundary characteristics in the Co-Cu alloys affect the emission of dislocations at the grain boundaries.

Molecular dynamics simulations

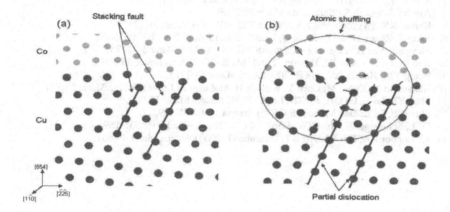

Figure 7. Atomic configuration in Co/Cu (554) grain boundary during the tensile test simulation under an applied stress of 2.0 GPa, (a) before dislocation emission (t=0 ps) and (b) after dislocation emission (t=5 ps).

MD simulations were performed in the nanocrystalline two-phase Co-Cu alloy to investigate the dislocation emission at Co/Cu interface. Figure 7 shows atomic configuration in Co/Cu (554) grain boundary before and after dislocation emission. Note that the stacking faults were generated before the tensile testing, as shown in Fig. 7 (a). This results from the intense geometrical strain at the Co/Cu interface. Partial dislocations were emitted to the Cu grain from the stacking faults. Dislocations were not emitted under an applied tensile stress of 1.7 GPa, but dislocations were emitted under an applied tensile stress of 2.0 GPa. Thus, a large stress was needed to emit the dislocations at the interface.

13

CONCLUSIONS

Three kinds of nanocrystalline Co-Cu alloys: a nanocrystalline Co-Cu alloy with nanoscale lamellar structure, a supersaturated solid solution Co-Cu alloy and a nanocrystalline two-phase Co-Cu alloy were processed by electrodeposition, and their mechanical properties were investigated at room temperature. These nanocrystalline Co-Cu alloys showed the high hardness and the low activation volume. The mechanical properties of the nanocrystalline Co-Cu alloys strongly depended on the grain boundary characteristics.

Molecular dynamics simulations revealed that the stacking faults, which are generated by the intense geometrical strain at the Co/Cu interface, play an important role in the dislocation emission at the interface.

REFERENCES

1. Wang N, Wang Z, Aust KT, Erb U (1997) Mater Sci Eng A237:150.
2. Iwasaki H, Higashi K, Nieh TG (2004) Scripta Mater 50:395.
3. Schiotz J, Tolla FDD, Jacobsen KW (1998) Nature 391:561.
4. Asaro RJ, Suresh S (2005) Acta Mater 53:3369.
5. Kumar KS, Swygenhoven HV, Suresh S (2003) Acta Mater 51:5743.
6. Jiang Z, Zhang H, Gu C, Jiang Q, Lian J (2008) J Appl Phys 104:053505.
7. Swygenhoven HV, Derlet PM, Frøseth AG (2006) Acta Mater 54:1975.
8. Dao M, Lu L, Asaro RJ, Hosson JTMD, Ma E (2007) Acta Mater 55:4041.
9. Wang YM, Hamza AV, Ma E (2006) Acta Mater 54:2715.
10. Nakamoto Y, Yuasa M, Chen Y, Kusuda H, Mabuchi M (2008) Scripta Mater 58:731.
11. Shen YF, Lu L, Lu QH, Jin ZH, Lu K (2005) Scripta Mater 52:989.
12. Shen YF, Lu L, Dao M, Suresh (2006) Scripta Mater 55:319.
13. Lu L, Schwaiger R, Shan ZW, Dao M, Lu K (2005) Acta Mater.53:2169.
14. Karimpoor AA, Erb U, Aust KT, Palumbo G (2003) Scripta Mater 49:651.

Mater. Res. Soc. Symp. Proc. Vol. 1224 © 2010 Materials Research Society   1224-FF08-03

Interface Effects on the Mechanical Properties of Nanocrystalline Nanolaminates

Alan F. Jankowski
Texas Tech University, Mechanical Engineering Department, Lubbock, TX 79409-1021, U.S.A.

ABSTRACT

Nanocrystalline nanolaminate (ncnl) structures are widely used in the study of physical properties in order to engineer materials for a variety of industrial applications. Often, novel and interesting mechanical behaviours that are found in nanolaminate materials can be linked with two characteristic features of structure. These are the layer pair spacing and the grain size. For the case of nanolaminates synthesized by physical vapor deposition processes, the layer spacing corresponds with the repeating sequence of layer pairs and can be referred to as composition wavelength. The grain size is the average width of the tapered columnar structure along the growth direction. Since the mechanical properties of strength and hardness are known to functionally vary with the separation between dislocations in crystalline materials, both structural features can potentially contribute to the total interfacial area and the characteristic separation of interfaces that mitigate dislocation motion. In this investigation, the individual contribution of layer pair spacing and grain size to the total interfacial structure are each quantified in an assessment of strength and hardness. A model is proposed for the total interfacial area of the material volume under plastic deformation that can quantify the interfacial area contribution from the layer pairs and the grain size. It is anticipated that each structural feature can potentially dominate the plastic deformation of the nanolaminate as dependent upon the specific layer pair spacing, the grain size, and the extent of plastic deformation.

INTRODUCTION

Synthesis and structure

Nanocrystalline nanolaminate (ncnl) structures are widely used in fundamental physical behaviour studies of materials. A variety of relevant applications are found industrially, such as optical band-pass filters for x-rays and neutrons [1-4], giant magneto-resistance [5-6] for high-density recording media, in low-temperature stability analysis [7-9], for bonding through high energetic reactivity and ultra-hard wear-resistant surfaces [10-13]. Novel and interesting mechanical behaviours can appear when the nanolaminate structures are artificially synthesized as, for example, through physical vapor or chemical deposition methods. In general, the physical properties are dependent upon the specific nanostructural arrangements of atoms at interfaces and grain boundaries as well as within the constituent layers.

Nanolaminates can have various degrees of crystallinity and are usually comprised of different elemental materials as, e.g., A and B in the A/B binary laminate system. The two primary structural features to describe the ncnl are: the characteristic layer pair spacing (h_λ) which is known as the composition wavelength for a repeating sequence of A and B layers; and the grain size (h_g). A schematic is shown in Fig. 1a for a typical ncnl structure wherein the grains assume a columnar type structure. Typical quantitative values for both h_λ and h_g are less than a 10^2 nm, and can be a small as 1 nm. For example, a bright-field transmission-electron micrograph of a gold-nickel (Au/Ni) ncnl viewed in cross-section is shown in Fig. 1b. The bright-field imaging condition reveals the composition modulation of the layer pairs along the growth direction (from bottom to top). In this case, the layer pair spacing h_λ equals 4.4±0.1 nm, and the grain size h_g equals 12±1 nm.

Often, ncnl(s) materials are prepared through one of several different physical vapor deposition methods, such as sputtering or thermal evaporation. In planar-magnetron sputter deposition, the cathodes are operated in the direct current and/or radio frequency mode to sputter a target surface with an ionized gas. The working gas is usually noble to avoid chemical interaction with the target material. Over a wide range of operation, the forward power applied to the cathode is proportional to the product of the discharge voltage and current. At lower power levels, typical deposition rates of 0.1-1.0 nm·sec^{-1} are found for many metals. The control of the residual stress state in a sputter deposited coating is accomplished by optimizing the working gas pressure. A transition in the residual stress state from compression to tension is reported [2] as the working gas pressure is increased from 0.4 to 4 Pa. The energetic parameters of the deposition process [14-15] are crucial to define the details of grain size, grain boundary structure, interface composition gradients, second phases, and lattice distortions.

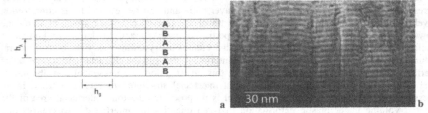

Figure 1 – The layer pair spacing (h_λ) and the grain size (h_g) shown in the (a) schematic are seen in the electron micrograph of (b) a Au/Ni nanocrystalline nanolaminate.

The use of x-ray diffraction in the θ/2θ mode with monochromatic radiation provides a method [16-21] to measure the crystalline texture, interplanar spacings, lattice distortions, layer pair spacing, composition profile, and grain size of the ncnl. In particular, the grain size (h_g) can be assessed from the width of the Bragg reflection that corresponds to the textured growth. Measurement of grain size is accomplished using the Debye-Scherrer formulation, i.e. h_g equals $c_B \cdot \lambda_{x\text{-ray}}/[B \cdot \cos(\theta)]$ where B is full-width at half-maximum intensity of the Bragg reflection, c_B is a constant (typically, 0.87-1.), and $\lambda_{x\text{-ray}}$ is the x-ray wavelength. The measured 2θ value for B is corrected for x-ray instrument broadening, and can be calibrated with results obtained by high magnification imaging of grain structure as accomplished with transmission electron microscopy. In general, the Debye-Scherrer method to determine grain size is quantitatively accurate over a range of 5-150 nm. The use of low-angle x-ray diffraction, i.e. at grazing incidence, provides an independent method to measure layer pair spacing, the ratio of the constituent layers, and the average interface roughness. The calculated reflectivity can be simulated [20] using the basic Fresnel equations. The analysis for determining the position and intensity of the satellite peaks about the Bragg reflections at high angle in θ/2θ scans provides a quantification of lattice distortions that are periodic with the layer pair spacing. The superlattice strain analysis is accomplished using kinematical and dynamical x-ray theory [16-18] to quantitatively distinguish the effects of an interfacial composition gradient, i.e. the composition profile, from the superlattice distortions. For metallic systems with mutual solubility, the superlattice distortions within the ncnl are found to relax towards unstrained values as the layer pair spacing increases above 5 nm as dependent upon the lattice misfit and deposition conditions. This type of strain relaxation is reported [13, 16-19] for metallic ncnl systems as Ta/V, Au/Ni and Cu/Ni. Additional features available for characterization include chemical bonding states, composition gradients, atomic and defect structure. Lattice distortions and measured relaxation from strained interfaces are evidenced in high resolution micrographs [22-23] where quantitative measurements of atomic positions can be used to assess x-ray diffraction results and simulations. Thus, characterization reveals the crystalline phases and unique lattice distortions of ncnl(s).

Strength of nanolaminates

Nanocrystalline metals can be characterized by two distinctly different structural regimes that consist of crystallites surrounded by grain boundaries [24] which often contain measurable disorder, i.e. excess free-volume. Strain rate effects on the measured strength [25-27] of several nanocrystalline metals detail this point. Consequently, the role of structural boundaries becomes more relevant as the grain size decreases, increasing its relative volume fraction. Thus, there comes a critical stage in grain size refinement [28-29] when microscopic deformation mechanisms, based on dislocation nucleation and mobility that explain strengthening via the classic Hall-Petch effect [30-36] may no longer apply. Several experimental studies [37-40] and accompanying structural simulations using molecular dynamics indicate diffusion regulated processes as grain boundary sliding accommodate deformation [41]. Therefore, the enhancement of strength that is found down to a critical refinement of grain size (or its distribution) [42-44] will then fade. Subsequently, strength values will fall away from the upper-bound ideal behavior. A present concern is to address how h_g and h_λ are relevant to interpret the mechanical properties of ncnl(s). In addition, the interface role between layers can be considered similar to solute (or precipitate) effects in generating lattice distortions and strain fields. The review of several ncnl systems is considered including face-centered-cubic (fcc) and body-centered-cubic (bcc) metals as well as ceramics that exhibit an array of diverse mechanical behaviors.

The mechanical testing of materials under uniaxial tension is the standard amongst methods to determine strength. Recent results [45] for macroscopically thick nanolaminates offer a glimpse into mechanical behavior trends for ductile two-phase materials wherein one of the repeat layers is amorphous to promote ductility. For fully crystalline ncnl(s) wherein all layers and interfaces between layers are crystalline, the use of tensile testing is a very challenging task. Some reliable results do exist. Micro-tensile testing [46] of 1 μm-thin sputter-deposited $Cu_{0.5}/Ni_{0.5}$ metal foils show an elastic-to-plastic transition marked by a distinct (upper and lower) yield point. A peak value of yield strength is found for a h_λ of 2.1 nm. The strain (ε_y) at the yield point (σ_y) was only 0.05-0.15% with stress-strain behavior characterized by unstable plastic flow. Fracture is seen at strain values less than 0.2%. A somewhat different behavior was reported [47] for evaporative deposits of $Cu_{0.50}/Ni_{0.47}Fe_{0.03}$ where a linear elastic regime from 0.04 to beyond 0.16% strain transitions into plastic behavior. As such, these evaporative deposits are suitable for an evaluation of plastic behavior wherein the yield strength is determined by the proportionality limit. In most ncnl cases, there is insufficient deformation to establish an offset yield point at 0.2% strain. Recently, an evaluation [48-49] of the variation of σ_y in Cu/NiFe ncnl(s) was conducted with respect to both h_g and h_λ. Local maxima in σ_y that exceeded 1 GPa values appear at h_λ of 2.4 and 4.2 nm, but there is no systematic trend found for the variation of σ_y with h_λ. On the other hand, the variation of σ_y versus $h_g^{-\frac{1}{2}}$ is seen to follow a Hall-Petch behavior with a strengthening factor (k_σ) of 4.56±0.79 GPa and an intercept value (σ_o) of 75±13 MPa for the intrinsic stress that is consistent with bulk alloy behavior. Therefore, in many

17

instances the deformation of ultra-high strength ncnl metals is often characterized by just a few-tenths of a percent of elastic strain. Often, little or no measurable plastic deformation follows, providing an uncertain measure of the yield strength. Since the onset of plastic behavior will provide a measure for the nanolaminate strength, the use of depth-sensitive indentation methods provides a consistent method to determine the role of nanoscale features on mechanical strength.

Hardness of nanolaminates

The role of coherency and shear modulus have been analyzed with respect to the plastic behavior of ncnl materials for systems that exhibit mechanical enhancements with variation in layer pair spacing. In a well-known approach [50], potential strength enhancements are assessed by considering the difference between the elastic moduli of constituent layers so thin that Frank-Read sources can not operate. The resolved shear stress that is then required to drive dislocations through the layers is on the order of $10^{-2} \cdot G$ where G is the shear modulus. This Koehler method has been applied to assess this energetic description to the origin of superhardness effects observed in ceramic nanolaminate systems, such as those [51-55] composed of various combinations of TiN, VN, NbN, TaN, AlN, and Al_2O_3 layers. In some cases, the difference in shear modulus was insufficient to account for the enhancements in hardness. For the case of the NbN/TaN ncnl(s) [52], the analysis of the elastic modulus difference only accounts for an enhancement of 7 GPa whereas a 20 GPa enhancement is measured. Contribution of coherency stress fields [52, 56-57] can then be used to account for the additional increase assuming that the misfit strain (ε) equals one-half the 3.2% lattice mismatch between $(111)_{NbN}$ and $(110)_{TaN}$. Also, coherency strain effects again account for the increases in hardness measured for AlN/VN [53].

For metallic ncnl(s), the basic Koehler analysis [50] is applied to quantify dislocation glide across interfaces with the resistance to shear. The yield stress (σ_y) can be estimated [58] as $\sigma_r \cdot m^{-1}$ from Schmid's Law, using a Taylor factor (m) of 0.32 for fcc metals and 0.30 for bcc metals. For the Cu/NiFe ncnl case, an estimate of the yield strength as 0.9-2.3 GPa is consistent with the range of experimentally measured values. For the evaluation of hardness in a Ta/V ncnl, an estimate from the yield stress as $3 \cdot \sigma_y$, i.e. $3 \cdot \sigma_r \cdot m^{-1}$, yields a hardness value of 5.3 GPa. In comparison to experimental findings for V/Ta [16], this value is much less than the enhancement that would be needed from the H_o value of 1.3 GPa to the 11.9 GPa value measured for a 3.14 nm V/Ta ncnl. Herein, coherency stress-field effects [13] are present in a strain-layered superlattice structure that can contribute up to an additional 14 GPa to the hardness assuming a 4.2% misfit strain, a 0.355 Poisson ratio, an 160 GPa average elastic modulus, and a composition modulation amplitude of 0.5. Thus, a variety of structural features and concurrent effects must be considered to fully account for potential enhancements in the strength and hardness of ncnl systems. Yet, it remains to be resolved the specific role that each dimensional feature (as grain size and layer spacing) have in determining the variations observed in strength and hardness.

METHODS

Experimental testing

The use of nanoprobe testing methods provides access to measurement conditions unattainable by macroscopic mechanical test methods. Loading to induce atomistic deformation in the plastic regime can be measured with control through nanoscratch test procedures which alleviates the constraint of depth sensitivity to 10 nm that is associated with nanoindentation methods. A Center for Tribology, Inc. NanoAnalyzer[TM] instrument is used to produce sets of nanoscratches in the surface of the nanolaminate coating with micron scale length and nanoscale width (w). These scratches are made with a cantilever mounted diamond stylus of known tip radius (r) using normal loads that range from 100 μN to 2 mN. The coating surfaces are scanned perpendicular to the nanoscratch to measure the width (and depth) at different positions along the scratch to compute average values. The scratch hardness can be computed by two methods. In the first, the nanoscratch width is compared to constant load nanoscratches produced on a reference material of known hardness. In the second method, the scratch hardness (H_s) is computed independently based [59] on the measured normal load (N) and scratch geometry as

$$N = \frac{1}{2} \cdot \pi \cdot H_s \cdot r^2 \cdot \sin(\alpha) \qquad (1)$$

where the contact angle (α) of the indenter tip with the surface is computed as

$$\alpha = \sin^{-1}[w \cdot (2r)^{-1}]. \qquad (2)$$

The force from the indenter tip is assumed to be loaded as a projection of the leading half of the indent along the direction of the scratch in eqn. (1). By substitution of eqn. (2) into (1), the scratch hardness (H_s) is

$$H_s = 8N \cdot (\pi \cdot w^2)^{-1}. \qquad (3)$$

An alternate expression to eqn. (3) for hardness can be computed using the exact area of the applied indenter load. For spherical tip contact where α is assumed equal to $\sin(\alpha)$, the scratch hardness equals

$$H_s = N \cdot (w \cdot r)^{-1}. \qquad (4)$$

Modeling contributions to interface separation

The dependence on grain size of mechanical behaviors including the plasticity of materials is of great interest since nanometer grain size structures offer high strengths and better wear resistance [59-61] as compared with conventional coarse grained materials. Strength enhancements attributed to high strain-rate sensitivity effects appear [53] to be governed by grain boundary deformation processes. Grain boundary sliding and rotation are intergranular deformation processes that are likely to dominate once grain size has been reduced to a point beyond which intragranular dislocation activity ceases. A strategy to make ultra-high strength nanomaterials [45, 62] is to limit the motion of dislocations required for plastic deformation. However, the ability to deform without failure is often reduced as a compromise to the ultra-high strength of most nanocrystalline metals. However, it may be possible to engineer the nanoscale structure of nanocrystalline nanolaminates for high strength and ductility at the same time.

The creation of interfaces to impede dislocation mobility and strengthen a solid are found in ncnl(s) in the form of grain boundaries and, possibly, the interfaces formed between nanolaminate layers as seen in Figs. 1a-b. The characteristic length scale (h_i) of the separation between these interfaces, as associated with the Hall-Petch relationship, is especially important to resolve and quantify when nanoscale probe techniques are used to measure the mechanical response. Changes may be anticipated in quantifying the characteristic interface separation length (h_i) when the probe shape and length are on the same scale as the specimen. Specifically, the depth of an indent can be selected to probe discrete layer pairs while sampling a large volume of grains. This is seen in the computation shown in Fig. 2 as a function of the indent width (w).

In the ncnl case of Fig. 2, a structure is modeled that has densely-packed, hexagonal-shaped columnar grains with a 8 nm grain size (h_g) and a 0.8 nm layer pair spacing (h_λ). The characteristic length (h_i) is determined as the ncnl is indented by a pyramidal-shaped Berkovitch stylus having a 50 nm tip radius (r). Each interface of the layer pair is assumed to contribute in this computation, so the asymptote value for deep indents equals $\frac{1}{2} \cdot h_\lambda$. For incoherent layer interfaces this assumption should be valid but this assumption may not always be the case. Some coherent interfaces may not act as effectively as obstacles to dislocation motion as dependent upon the resultant stress-strain field. Recall that many ncnl(s) show a transition from coherent to incoherent interfaces as the layer pair spacing changes. Modification to the characteristic h_λ should be considered to accommodate this transition of interface state.

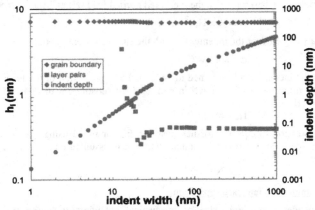

Figure 2 – The characteristic length (h_i) that separates internal boundaries as computed for a nanocrystalline nanolaminate with a 8 nm grain size and 0.8 nm layer pair spacing.

The indent depth can be computed as a function of its width (w) during indentation. A transition occurs with increasing indent depth from sensing grain boundaries alone to inclusion of layer pair interfaces. This transition is found in Fig. 2 for the 8 nm grain size (h_g) and a 0.8 nm layer pair spacing (h_λ) ncnl for indentation widths of 10-50 nm, and a corresponding depth of 1-5 nm. A second transition effect is computed to occur at a 20 nm indent width when the indent shape changes as the stylus penetrates into the specimen beyond the tip radius.

Contributions to a characteristic length (h_i) can be modeled as the sum contribution of layer pair interfaces and grain boundaries by computation of a characteristic volume to interfacial area ratio. (However, the relative role in contributing to strength and ductility remains unresolved at this point in examination of the mode and mechanism for deformation.) For the case of experimental testing with indenter tips, the characteristic separation between dislocations (h_i) can be equated to the total volume of the material plastically deformed by the indenter displacement (V_i) divided by the total interfacial area (A_i) within that displacement volume. Since this characteristic separation will change with depth, the mechanical properties measured are likely to represent that change in h_i. The proportionate roles of grain size (h_g) and layer pair spacing (h_λ) at the onset of plasticity, i.e. the yield strength as well as the hardness, can be quantified and compared. In addition, the deformation and ensuing strength of the ncnl may be considered a consequence of a possible two-stage deformation process that results from each characteristic structural feature. An asymptotic value for the characteristic length between dislocations can be computed for h_i considering contributions from both layer pair spacing (h_λ) and grain size (h_g). A first computation of the averaged interface separation is made using a cylindrical crystalline volume ($V_{\lambda g}$) formed by the grain size and layer-pair interface spacings that is equated to a characteristic spherical volume (V) with the diameter ($h_{\lambda g}$) as

$$h_{\lambda g} = [(\tfrac{3}{4}) \cdot (h_g)^2 \cdot (h_\lambda)]^{\frac{1}{3}} \tag{5}$$

assuming

$$V_{\lambda g} = (\pi/8) \cdot (h_g)^2 \cdot (h_\lambda), \tag{6a}$$

$$V = (\pi/6) \cdot (h_{\lambda g})^3, \tag{6b}$$

and

$$V_{\lambda g} = V. \tag{6c}$$

The height of the cylindrical crystalline volume ($V_{\lambda g}$) is assumed as one-half the layer pair (provided each laminate interface is considered a barrier to plastic deformation), and the width as equivalent to the columnar width as represented by in-plane measurements of grain size. The assumption for height is made but can be debated for at least two reasons. The first reason regards the coherent-incoherent nature of the lattice structure along the growth direction. That is, if the laminate interfaces are not incoherent, then are the resultant stress-strain fields produced sufficient to repel dislocation motion. This assumption is often added as a contribution of coherency effects [48-57] when evaluating hardness based on the Koehler model. The second reason regards whether or not dislocations actually move toward the laminate interfaces. For example, edge dislocations that are seen [22-23] in high resolution images of cross-sectioned Au/Ni ncnl(s) would glide parallel to the interfaces at the onset of plastic deformation.

RESULTS

Interface contributions to hardness

The strengthening of bulk polycrystalline nanostructures by the Hall-Petch mechanism [30-36] is well established, and can be empirically derived [37] as inversely proportional to the square-root of a dimensional feature (h_i) that represents characteristic separation of dislocations. The tensile strength (σ) is often expressed as

$$\sigma = \sigma_o + k_\sigma \cdot (h_i)^{-\frac{1}{2}} \tag{7}$$

where σ_o is the intrinsic strength, and k_σ the strengthening coefficient. Similarly, if hardness is considered a measure of strength [35, 63] at the onset of plastic deformation, then a Hall-Petch expression follows as

$$H = H_o + k_H \cdot (h_i)^{-\frac{1}{2}} \tag{8}$$

where H_o is the intrinsic hardness of the material without (e.g. grain boundary) hardening, and k_H the hardening coefficient.

A recent review [60] describes the mechanical behavior of nanocrystalline grain size for several cubic metals. In general, at some degree of grain refinement to the nanoscale, the enhancement of strength becomes saturated and then softens as the constituent nanocrystalline phase eventually transitions towards an amorphous structure. The anticipated loss of strength for nanocrystalline materials with the finest of grain size coincides with the absence of conventional dislocation-based mechanisms [64-65], and is generally described as evolving into a transition from perfect dislocation slip into partial dislocation assisted twinning and stacking faults [66], followed by a variant of grain boundary migration [67] with triple-junction motion [68-69] as associated to the case of high strain-rate plasticity. For this regime of nanoscale deformation, mechanical softening can be modeled with molecular dynamics using mechanism(s) of grain

21

boundary sliding. Recently, a multi-scale modeling effort [70] similarly describes the entire spectrum of mechanical strengthening through softening using a three-phase material equivalent. In these approaches, the localization of plastic deformation prior to fracture produces grain boundary morphologies similar to the intrinsic structure of ncnl(s) as illustrated in the Figs. 1a-b, i.e. regions of deformation having near parallel grain boundaries. Thus, variations in the mechanical properties are anticipated, observed, and depend upon the nanoscale features.

Results for nanoindentation hardness measurements of 400-500 nm thick Au/Ni ncnl(s) are shown [48, 71] in the Fig. 3 Hall-Petch plots using eqn. (8) for specimens that were sputter deposited onto Si wafers with an epitaxial growth layer of Au (or an adhesion layer of Cr) at room temperature using a 0.67 Pa working gas pressure of Argon. The applied load force and tip displacement are continuously measured during the process and the data reduction is conducted as described [72] by the Oliver-Pharr method. Hardness data is considered for indentation depths less than 20% the film thickness for reasons to be clarified later. In addition, Vickers microhardness results [73] are plotted for electrodeposits of nanocrystalline Ni using a 50 gf indent load. The grain size of the nanocrystalline Ni samples was determined using the lineal intercept method [74] accounting for both twin and large angle grain boundaries.

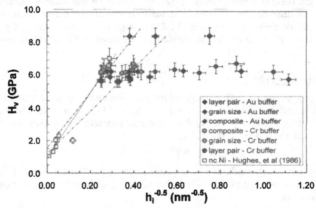

Figure 3 – The nanoindentation hardness (H_v) of Au/Ni ncnl(s) as a function of the layer pair spacing and grain size along with Hughes, et al. (1986) data for nanocrystalline Ni.

The variation of hardness (H) is shown in Fig. 3 with layer pair spacing (h_λ), grain size (h_g), and a composite interface separation ($h_{\lambda g}$) as computed using eqn. (5) where attributed to both layer pair spacing and grain size. The grain size is determined from the broadening of the (111) Bragg reflection, and the layer pair spacing is measured from short-range order peaks about the (111) Bragg reflections for these Au/Ni ncnl coatings that have a (111) growth texture but are polycrystalline in-plane [22-23]. The values for H_o and k_H computed using a linear regression fit for the data with each h_i is listed in Table 1. The k_H and H_o values for the grain size (h_g) and composite interface spacing ($h_{\lambda g}$) curve fits of the Au/Ni ncnl(s) are consistent with the electrodeposited nanocrystalline Ni results. The curve fits based on the layer pair spacing (h_λ) alone, yields a H_o value that is not consistent with known physical behavior. From this limited sampling, it can not be clearly distinguished whether the composite interface spacing ($h_{\lambda g}$) or grain size (h_g) provides the best fit, although the H_o value for composite value appears more consistent with measured behavior for pure Ni where H_o is less than or equal to ~1 GPa.

Table I. Hall-Petch coefficients (k_H; H_o) as a function of interface separation (h_i)				
Coefficient	Au/Ni nanocrystalline nanolaminates			Nanocrystalline Ni
	h_λ (>1.8 nm)	$h_{\lambda g}$	h_g	
k_H (GPa·nm$^{-1/2}$)	3.6 ± 2.1	13.4 ± 4.1	16.6 ± 4.5	20.6 ± 1.0
H_o (GPa)	4.29	1.10	1.79	0.79

Indentation effects

The mechanical properties of the Au/Ni ncnl(s) are revisited considering the nanoscale features that need to be accounted for when analyzing hardness. These are the nanoindentation depth as well as the contributions from the layer pair interfaces and the grain boundaries. The Fig. 4 plot utilizes the layer pair spacing (h_λ) as the characteristic length parameter.

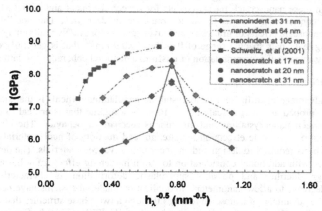

Figure 4 – The effective depth of nanoindentation and nanoscratch tests affect the hardness (H) measurement of Au/Ni ncnl(s) are seen in these Hall-Petch plots as a function of layer pair spacing (h_λ) along with data reported by Schweitz, et al. (2001).

The hardness variation shows an increase with decreasing layer pair spacing and a peak value near a 2 nm layer pair. Some observations [21] suggest that the nanoindentation hardness increases as the layer pair spacing decreases, and that hardness is not dependent on the depth of the indent over a indentation load range of 1-8 mN. However, to thoroughly evaluate the nanoindentation hardness, the indentation depth must be considered since it is shown [75] that the substrate can increasingly influence the measured hardness as the indentation depth increases. For Au/Ni ncnl(s) on Si substrates, this somewhat abrupt transition from sensing the coating to sensing the substrate is reported [71] at depths greater than 40-50 nm using a Meyer plot of indentation depth versus load. The measured hardness values will eventually approach the 9.9 GPa value of the silicon substrate wafer [76] with increased indentation depth.

In addition, an increase in the hardness of face-centered-cubic (fcc) materials should occur with an increase in the plastic deformation as achieved at greater indentation loads and indentation depths. This gradual increase in hardness can occur since fcc ncnl(s) metals are known to work harden as revealed [47] in uniaxial tensile tests of Cu/Ni. This effect of indentation depth on hardness may appear for the samples with layer pair spacings greater than 2 nm, i.e. $h_\lambda^{-0.5} < 0.7$ nm$^{-0.5}$, where the layer pair interface structure is somewhat invariant as characterized [22-23] by twinned Au-on-Ni interfaces and edge dislocations: within the Ni layers [48] that accommodate the lattice misfit between the Au and Ni; terminate at each Au-Ni interface; and glide when mobilized parallel to the interfaces. Hardening with an increase in plastic volume due to increasing indent depth may result as well for the Au/Ni ncnl(s) with layer pair spacings less than 1.2 nm, i.e. $h_\lambda^{-0.5} > 0.9$ nm$^{-0.5}$, wherein the layer pair interface structure is again somewhat invariant as characterized in this range by coherent interfaces.

The nanoscratch hardness (H$_s$) results as computed using eqn. (3) for nanoscratch tests are added to the Hall-Petch plot of Fig. 4 for the Au/Ni ncnl sample with a 6.9 nm grain size and 1.8 nm layer pair spacing at scratch depths of 17, 20, and 31 nm. The increase in nanoindentation hardness previously observed at the most shallow, effective indentation depth of 31 nm (a trend now opposite the work hardening effect for an invariant interface structure) is further accentuated in these nanoscratch measurements for scratch depths at and below 31 nm wherein the hardness continues to rise as the nanoscratch depth decreases. The possible interaction between the nanoscale effects of grain size and layer spacing for Au/Ni ncnl(s) may be evident in this remaining layer pair spacing range of $0.7 < h_\lambda^{-0.5} < 0.9$ nm$^{-0.5}$ that is generally characterized [18, 22-23] by a near full accommodation of misfit in a distorted coherent superlattice structure.

DISCUSSION

The advantage of utilizing a nanocrystalline nanolaminate (ncnl) is that size effects can be selectively probed at a length scale below 10 nm – a regime that is critical to exploring the transition between nanocrystalline and amorphous mechanical behavior. The characterization and assessment of nanoscale effects on the strength and ductility of ncnl materials can provide the scientific basis required to design and engineer new advanced materials. The development of ncnl structures with additional consideration to strain hardening effects may be quite useful in obtaining high ductility and possibly superplastic deformation as characteristic interface separations decrease to a few nanometers. In application, material systems have been developed in order to take advantage of nanoscale features through a two-phase structure that displays two-stage deformation. For the case of nanostructured steels [77], a ductile fine-grained matrix provides high strength and ductility whereas the second phase, i.e. the nanoscale precipitate, provides further strengthening as these hard nanoscale features affect the work hardening regime of plastic deformation. For the ncnl case, the effects of nanoscale features as grain size and layer spacing may too provide a path for designing a two-phase nanostructured composite. The initial yielding could be attributed to the effect of grain size, for example, and the layer spacing as a second dimensional variant may then contribute to the work hardening regime to extend ductility. That is, whereas grain size may be tuned to provide the initial yield point, the laminate spacing could provide subsequent influence on the hardening behavior during plastic deformation to ultimate strength with potential incorporation of coherency effects.

CONCLUSIONS

The separation of interfaces as determined by grain boundaries, for example, influences the mechanical behavior of nanocrystalline materials. In nanocrystalline nanolaminate (ncnl) structures, the separation of interfaces takes form in consideration to the contributions of layer pair interfaces and grain boundaries. An apparent softening in the Hall-Petch effect for ncnl hardness below a few nanometers when considering layer pair spacing alone is abated through use of a composite interface-separation parameter. The experimental trends observed for the hardness variation of Au/Ni ncnl(s) with indentation depth can be accounted for using a composite interface separation parameter that includes the contributions of both layer separation and grain size to define a characteristic volume for the separation of dislocations.

ACKNOWLEDGMENTS

Support for this work was provided through the J.W. Wright Endowment for Mechanical Engineering at Texas Tech University.

REFERENCES

1. A.F. Jankowski, D.M. Makowiecki, et al., J. Appl. Phys. 65 (1989) 4450-4451
2. A.F. Jankowski, R.M. Bionta and P.C. Gabriele, J. Vac. Sci. Technol. A 7 (1989) 210-213
3. A.F. Jankowski, Optical Eng. 29 (1990) 968-972
4. A.F. Jankowski, SPIE Conf. Proc. 1738 (1992) 10-21
5. J.R. Childress, C.L. Chien and A.F. Jankowski, Phys. Rev. B 45 (1992) 2855-2862
6. A. Simopoulos, E. Devlin, et al., Phys. Rev. B 54 (1996) 9931-9941
7. A.F. Jankowski and T. Tsakalakos, Metall. Trans. A 20 (1989) 357-362
8. A.F. Jankowski and C.K. Saw, Scripta Mater. 51 (2004) 119-124
9. A.F. Jankowski, Defect and Diffusion Forum 266 (2007) 13-28
10. D.M. Makowiecki, A.F. Jankowski, et al., J. Vac. Sci. Technol. A 8 (1990) 3910-3913
11. A.F. Jankowski, M.A. Wall, et al., NanoStructured Mater. 9 (1997) 467-471
12. A.F. Jankowski, J.P. Hayes, et al., Thin Solid Films 308/309 (1997) 94-100
13. A.F. Jankowski, J.P. Hayes, and C.K. Saw, Phil. Mag. 87 (2007) 2323-2334
14. A.F. Jankowski, Thin Solid Films 220 (1992) 166-171
15. A. Jankowski, J. Hayes, J. Nilsen, et al., Thin Solid Films 469-470 (2004) 372-376
16. S. Jayakody, J. Chaudhuri and A.F. Jankowski, J. Mater. Sci. 32 (1997) 2605-2609
17. J. Chaudhuri, S.M. Alyan and A.F. Jankowski, Thin Solid Films 219 (1992) 63-68
18. J. Chaudhuri, S. Shah, et al., J. Appl. Phys. 71 (1992) 3816-3820
19. A.F. Jankowski, Superlattices and Microstructures 6 (1989) 427-429
20. V.G. Kohn, Physica Status Solidi (b) 187 (1995) 61-70
21. K.O Schweitz, J. Chevallier, J. Bottiger, et al., Phil. Mag. A 81 (2001) 2021-2032
22. A.F. Jankowski, J. Appl. Phys. 71 (1992) 1782-1789
23. M.A. Wall and A.F. Jankowski, Thin Solid Films 181 (1989) 313-321
24. Y.M. Wang and E. Ma, Applied Physics Letters 85 (2004) 2750-2752
25. Q. Wei, S. Cheng, K.T. Ramesh, and E. Ma, Mater. Sci. and Eng. A 381 (2004) 71-79
26. R.P. Carreker and W.R. Hibbard, Acta Metallurgica 1 (1953) 656-658
27. L. Lu, S.X. Li, K. Lu, Scripta Materialia 45 (2001) 1163-1169
28. J.R. Weertman, in Nanostructured Materials, C. Koch (ed.), William Andrew Press, Norwich (2007) pp 537-564
29. M. Chen, E. Ma, and K. Hemker, in NanoMaterials Handbook, Y. Gogotsi (ed.), CRC Press, New York (2006) pp 407-529
30. E.O. Hall, Proc. Physical Society B 64, (1951) 747-753
31. E.O. Hall, Journal of Mechanics and Physics of Solids 1 (1953) 227-233
32. N.J. Petch, Journal Iron Steel Institute 174 (1953) 25-28

33. N.J. Petch, Progress in Metal Physics 5 (1954) 1-10
34. N.J. Petch, Acta Metallurgica 12 (1964) 59-65
35. R.M. Douthwaite and N.J. Petch, Acta Metallurgica 18 (1970) 211-216
36. N.J. Petch and R.W. Armstrong, Acta Metallurgica Materialia 38 (1990) 2695-2700
37. A.H Chokshi, A. Rosen, J. Karch, H. Gleiter, Scripta Metallurgica 23 (1989) 1679-1683
38. R. Schwaiger, B. Moser, et al., Acta Materialia 51 (2003) 5159-5172
39. C. Schuh, T.G. Nieh, T. Yamasaki, Scripta Materialia 46 (2002) 735-740
40. C. Schuh, T.G. Nieh, H. Iwasaki, Acta Materialia 51 (2003) 431-443
41. H. Van Swygenhoven, D. Farkas, and A. Caro, Phys. Rev. B 62 (2000) 831-838
42. T. Ungár, S. Ott, P. Sanders, A. Borbély, and J. Weertman, Acta Mater. 46 (1998) 3693-3699
43. M. Ke, S. Hackney, W. Milligan, and E.C. Aifantis, Nanostructured Mater. 5 (1995) 689-697
44. Z.W. Shan, E.A. Stach, et al., Science 305 (2004) 654-657
45. Y.M. Wang, J. Li, et al., Jr., Proc. National Acad. Sci. U.S.A. 104 (2007) 11155-11160
46. A. Jankowski and T. Tsakalakos, J. Applied Physics 57 (1985) 1835-1838
47. A.F. Jankowski, E.M. Sedillo and J.P. Hayes, Japan. J. Appl. Phys. 33 (1994) 5019-5025
48. A.F. Jankowski, Surface and Coatings Technology 203 (2008) 484-489
49. A.F. Jankowski and J.F. Shewbridge, Materials Letters 4 (1986) 313-315
50. J.S. Koehler, Physical Review B 2 (1970) 547-551
51. X. Chu and S.A. Barnett, J. Applied Physics 77 (1995) 4403-4411
52. J. Xu, M. Kamiko, et al., J. Applied Physics 89 (2001) 3674-3678
53. G. Li, J. Lao, J. Tian, Z. Han, and M. Gu, J. Applied Physics 95 (2004) 92-96
54. M. Ben Daia, P. Aubert, S. Labdi, et al., J. Applied Physics 87 (2000) 7753-7757
55. L. Wei, M. Kong, Y. Dong, and G. Li, J. Applied Physics 98 (2005) 074302-1-4
56. J.W. Cahn, Acta Metallurgica 11 (1963) 1275-1282
57. M. Kato, T. Mori, and L.H. Schwartz, Acta Metallurgica 28 (1980) 285-290
58. G.I. Taylor, Journal of the Institute of Metals, 62 (1938) 307-324
59. K. M. Lee, C. D. Yeo, A. A. Polycarpou, Experimental Mechanics 47 (2007) 107-121
60. M. Dao, L. Lu, R.J. Asaro, J.T.M. De Hosson, and E. Ma, Acta Mater. 55 (2007) 4041-4065
61. C. D. Gu, J. S. Lian, Q. Jiang, W.T. Zheng, J. Physics D: Appl. Phys. 40 (2007) 7440-7446
62. T. Zhu, J. Li, A. Samanta, et al., Proc. National Acad. Sci. U.S.A. 104 (2007) 3031-3036
63. J.T. Burwell and C.D. Strang, Proc. Roy. Soc. Lond. A, Math. Phys. Sci. 212 (1952) 470-477
64. C.S. Pande, R.A. Masumura, and R.W. Armstrong, Nanostructured Mater. 2 (1993) 323-331
65. T.G. Nieh and J. Wadsworth, Scripta Metallurgica et Materialia 25 (1991) 955-958
66. M. Chen, E. Ma, and K. Hemker, in Nanomaterials Handbook, Y. Gogotsi (ed.), Taylor and Francis, Boca Raton (2006) p. 523
67. H. Hahn, P. Mondal, and K.A. Padmanabhan, Nanostructured Materials 9 (1997) 603-606
68. K.S. Kumar, S. Suresh, M.F. Chisholm, et al., Acta Mater. 51 (2003) 387-405
69. M. Yu. Gutkin, I.A. Ovid'ko, and N.V. Skiba, Acta Materialia 52 (2004) 1711-1720
70. S. Benkassem, L. Capolungo, and M. Cherkaoui, Acta Materialia 55 (2007) 3563-3572
71. A.F. Jankowski, J. Magnetism and Magnetic Materials 126 (1993) 185-191
72. W.C. Oliver and G.M. Pharr, J. Materials Research 7 (1992) 1564-1583
73. G.D. Hughes, S.D. Smith, C.S. Pande, et al., Scripta Metallurgica 20 (1986) 93-97
74. R.C. Gifkins, Optical Microscopy of Metals, Elsevier, New York (1970) p. 178
75. C. Feldman, F. Ordway, and J. Bernstein, J. Vac. Sci. Technol. A 8 (1990) 117-122
76. A.A. Giardini, The American Mineralogist 43 (1958) 957-969
77. J.W. Morris, Jr., Inter. Offshore Polar Engineering Conf. Proc. 16 (2007) 2814-2818

Mater. Res. Soc. Symp. Proc. Vol. 1224 © 2010 Materials Research Society   1224-FF05-22

RESIDUAL STRESS REDUCTION IN SPUTTER DEPOSITED THIN FILMS BY DENSITY MODULATION

Arif S. Alagoz[1], Jan-Dirk Kamminga[2,3], Sergey Yu. Grachev[4], Toh-Ming Lu[5] and Tansel Karabacak[1]

[1] Department of Applied Science, University of Arkansas at Little Rock, Little Rock, AR 72204, U.S.A.
[2] Materials Innovation Institute M2i, Mekelweg 2, 2628 CD Delft, Netherlands
[3] Department of Materials Sciences and Engineering, Delft University of Technology, Mekelweg 2, 2628 CD Delft, Netherlands
[4] Saint-Gobain Recherche, 39 quai Lucien Lefranc, B.P. 135, 93303 Aubervilliers, France
[5] Department of Physics, Applied Physics and Astronomy, Rensselaer Polytechnic Institute, Troy, NY 12180, U.S.A.

ABSTRACT

Control of residual stress in thin films is critical in obtaining high mechanical quality coatings without cracking, buckling, or delamination. In this work, we present a simple and effective method of residual stress reduction in sputter deposited thin films by stacking low and high material density layers of the same material. This multilayer density modulated film is formed by successively changing working gas pressure between high and low values, which results in columnar nanostructured and dense continuous layers, respectively. In order to investigate the evolution of residual stress in density modulated thin films, we deposited ruthenium (Ru) films using a DC magnetron sputtering system at alternating argon (Ar) pressures of 20 and 2 mTorr. Wafer's radius of curvature was measured to calculate the intrinsic thin film stress of multilayer Ru coatings as a function of total film thickness by changing the number of high density and low density layers. By engineering the film density, we were able to reduce film stress more than one order of magnitude compared to the conventional dense films produced at low working gas pressures. Due to their low stress and enhanced mechanical stability, we were able to grow these density modulated films to much higher thicknesses without suffering from buckling. Morphology and crystal structure of the thin films were investigated by scanning electron microscopy (SEM), transmission electron microscopy (TEM), and X-ray diffraction (XRD). A previously proposed model for stress reduction by means of relatively rough and compliant sublayers was used to explain the unusually low stress in the specimens investigated.

INTRODUCTION

It is well known that residual (also called intrinsic) stress naturally evolves during growth of sputter deposited thin films. The control of this stress is essential for high mechanical quality coatings without cracking, buckling, or delamination [1-2]. Dependence of residual stress on process parameters and microstructure is systematically investigated by many research groups. Haghiri-Gosnet et al. [3] and Windischmann [4] showed strong correlation between residual stress and Thornton's structure zone model (SZM) [5-9] for sputtered thin films. Low density thin films deposited at high argon pressure and low temperature (regime Zone I) have almost zero residual stress. However, these films exhibit poor electrical properties due to their columnar microstructure. Decreasing argon pressure develops tensile stress and further decrease in pressure results in a dramatic change from tensile to compressive stress finally forming a dense thin film (regime Zone T). In order to deposit low stress thin films, setting working pressure between tensile to compressive stress transition region is not feasible since the working pressure window is too narrow and the deposition is not stable. Karabacak et al. [10] demonstrated that compressive stress build-up in a film can be reduced by modulating between relatively rough and compliant sublayers, and dense sublayers. Overall thin film stress was reduced in sputter deposited tungsten thin films by successively changing sputtering conditions leading to a density modulated multilayer thin film composed of low and high density layers. The low density layer consisted of nano-columns deposited under high pressure is believed to be acting as compliant layer. It was proposed that a rough compliant underlayer delays the development of compressive stress in the following dense layer over a thickness in the order of the underlayer's roughness.

In this paper we present stress reduction in multilayer ruthenium (Ru) sputtered thin films of more than one order of magnitude compared to the conventional dense films by engineering the film density. It will be shown that the low density films exhibit a significant roughness and are likely to be acting as a compliant layer, thus explaining the low stresses observed. It is also shown that the deposition pressure has an important effect on crystal orientation, where high and low pressure depositions results in a different crystal texturing in density modulated thin films.

EXPERIMENT

All Ruthenium depositions are performed at room temperature by dc magnetron sputtering technique. Films were deposited on n-type, 12-25 $\Omega \cdot$cm resistivity, (100) oriented single crystal silicon wafers using 99.95% pure 3 inch ruthenium target. All samples were mounted parallel to target at 18 cm distance. Before each deposition, 5×10^{-7} Torr base pressure was achieved by a turbo-molecular pump. All depositions were performed at 200 W DC power using ultra pure Argon plasma. Argon pressure was set to 2 mTorr in order to deposit high density single layer films and 20 mTorr for low density columnar single layer thin films. Density modulated multilayer thin film depositions started with 15 nm thick low density film grown at 20 mTorr followed by 17 nm dense film at 2 mTorr. Pressure level was successively changed between these two values until the desired thin film thickness is reached.

Intrinsic thin film stress was calculated from the Stoney equation [11]. Radius of curvature of wafer before and after the deposition was measured by using a Flexus 2320 dual wavelength ($\lambda_1 = 670$ nm and $\lambda_2 = 750$ nm) system. Thin films' morphology and thickness were investigated by using Carl Zeiss Ultra 1540 dual beam focused ion beam (FIB)/scanning electron microscopy (SEM) system. Crystal structure of the films was studied by using a Scintag X-ray diffractometer (XRD) using a Cu target operated at 50 kV and 30 mA. Transmission electron microscopy (TEM) was performed on cross sections of Ru multilayers, prepared by ion milling in a Gatan PIPS 691 ion mill, using Ar. Specimens were analyzed in a CM30T Philips transmission electron microscope operated at 300 kV.

DISCUSSION

Figure 1 shows the measured stress values of single layer dense ruthenium thin films deposited at low argon pressure, low density films deposited high argon pressure, and multilayer thin films deposited at successive high/low argon pressure as a function of total film thickness.

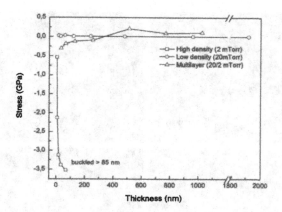

Figure 1. Stress evolution as a function of film thickness.

During deposition, a highly compressive residual stress develops at the Ru thin films deposited at a low argon pressure. These high material density films could not be grown thicker than 85 nm, after which they peeled off from the substrate. In contrast, we could deposited almost 2 μm thick low density single layer thin films at high argon pressure with close to zero tensile residual stress. SEM images show that these films exhibit a columnar microstructure as predicted by the structure zone model (Fig. 2 a-b). Deposition at high argon pressure reduces the directionality of the incoming ruthenium flux to the substrate and enhances the shadowing effect during film deposition which is responsible for the formation of a columnar structure [12]. Multilayer Ru thin films deposited by changing argon pressure successively between high and low pressure are relatively dense (Fig. 2 c-d) and show very low residual stress values (Fig. 1). This way, we reduced total residual stress of ruthenium films one order of magnitude compared to the conventional dense films. Using density modulation technique, very thick films with reasonably low stresses can be prepared. Interestingly, unlike the single layer high density Ru films, the stress of multilayer film increases slowly as the total thickness increases. This tendency can be explained by thermal stress evolution during long deposition times due to the thermal expansion coefficient mismatch between Si substrate (2.6×10^{-6} K^{-1}) and Ru (6.4×10^{-6} K^{-1}) film. Thermal stress seems to stabilize at thicker samples (Fig. 1).

The microstucture of multilayer thin films was further examined by TEM. In Fig. 3, the TEM image of a density modulated layer clearly shows high and low density sublayers, where the low density layers exhibit a marked porosity. According to Ref. [10] stress build-up in dense films is delayed by a compliant rough layer as long as the dense layer thickness is of the order of the roughness of the interface. Figure 3 shows that the interface is sufficiently rough. The porosity of the low density layer has a nanocrack-like appearance which suggests that the layer is also sufficiently compliant to reduce the stress in the dense layer. (On the basis of the present results we cannot conclude whether the nanocrack-like features were produced by deformation, or are a direct result of the growth process).

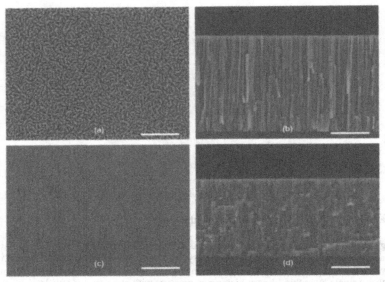

Figure 2. Top and cross-sectional view SEM images of (a-b) low density and (c-d) low/high density multilayer samples. Scale bars at each image measure 500 nm.

Consequently, the findings are in good correspondence with the model proposed in [10]. In Ref. [13] an alternative model for stress generation was proposed, in which the total stress is the sum of a compressive growth stress, and a tensile stress created at grain boundaries. This model could also explain the current results, because the current deposition strategy leads to small grains. However, the model in Ref. [13] relies on the columnar microstructure of the layers. In the present layers, a clear columnar microstructure is not observed. Therefore we attribute the low stress to the compliance of the low density sublayers.

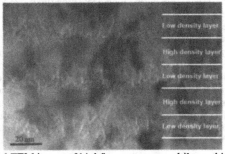

Figure 3. Cross-sectional TEM image of high/low pressure multilayer thin film.

In addition, we investigated the crystal orientation (texture) of thin films by XRD and observed that crystal orientation of the grains is strongly affected by argon pressure level. As shown in Fig. 4, thin films deposited at low pressure are textured along (002) and (101) orientations. On the other hand, thin films deposited at high pressure are textured along the (100) orientation. Different texture evolution for different argon pressure levels can be explained by the increased shadowing effect at a high pressure and different surface mobility of adatoms on different crystal orientations. Karabacak [14, 15] presented that at the early stages of deposition, low surface mobility islands can grow vertically as opposed to higher surface mobility islands which prefer lateral island growth. Due to their taller height, the shadowing effect enhances further vertical growth of low adatom mobility islands making them the grains of dominant crystal orientation. In a similar way, increasing argon pressure enhances the shadowing effect and the crystal growth on low mobility crystal orientation. XRD measurements of multilayer thin films exhibit the crystal orientation contribution of both low and high density layers. However, the profile of the density modulated film is shown to be quite different than a simple composition of the profiles of each layer. In the density modulated film (002) orientation becomes quite insignificant unlike in the single layer high density films. This can be attributed to a "resetting" effect, in which just before the (002) islands start to grow at the expense of (101) islands during the low pressure step, Ar pressure is set to high and the (100) island growth is promoted. When the pressure is set to low again, the process starts from the beginning again resetting the growth. This mechanism is also believed to influence the dynamic evolution of the island size. By resetting the growth at each layer, we keep the average grain size small before it has a chance to grow. We are currently investigating this property in more detail.

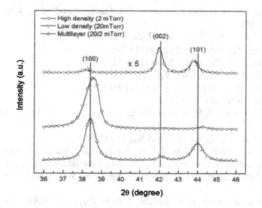

Figure 4. Crystal orientation (texture) of Ru films.

We also note that, in Fig. 4, central peak positions in the XRD profiles of a single layer high and low density films are shifted from their equilibrium positions, which is due to the residual stress [16]. On the other hand, there is no such a shift in low stress density modulated thin films of Ru.

CONCLUSIONS

In conclusion, it has been shown that material density of a sputter deposited film can be engineered by changing the argon pressure between high and low levels. This approach results in a density modulated multilayer thin film having high and low density layers produced by low and high working gas pressures, respectively. By this way, the residual compressive stress values can be reduced more than one order of magnitude and the requirement of a critical thickness for films without buckling may be relaxed. The low density layers act as a compliant layer to reduce the stress of the subsequent deposited high density layer. In addition, the crystal orientation profile of density modulated films is shown to be quite different than a simple composition of the profiles of each layer.

ACKNOWLEDGEMENTS

F.D Tichelaar at the National Centre for HREM at Delft University of Technology is acknowledged for the investigations by transmission electron microscopy. This work was supported by the NSF.

REFERENCES

1. *Thin Film Materials: Stress, Defect Formation and Surface Evolution*, L. B. Freund and S. Suresh, (Cambridge University Press, 2004)
2. D. C. Meyer, A. Klingner, T. Holz, and P. Paufler, *Appl. Phys. A: Mater. Sci. Process.* **69**, 657 (1999).
3. A. M. Haghiri-Gosnet, F. R. Ladan, C. Mayeux, H. Launois, and M. C. Joncour, *J. Vac. Sci. Technol. A* **7**, 2663 (1989).
4. H. Windischmann, *J. Vac. Sci. Technol. A* **9**, 2431 (1991).
5. J. A. Thornton, *J. Vac. Sci. Technol.* **11**, 666 (1974).
6. J. A. Thornton, *J. Vac. Sci. Technol.* **12**, 830 (1975).
7. J. A. Thornton, *Annu. Rev. Mater. Sci.* **7**, 239 (1977).
8. J. A. Thornton, *J. Vac. Sci. Technol. A* **4**, 3059 (1986).
9. J. A. Thornton, *Proc. SPIE* **821**, 95 (1987).
10. T. Karabacak, C. R. Picu, J. J. Senkevich, G.-C. Wang, and T.-M. Lu, *J. Appl. Phys.* **96**, 5740 (2004).
11. G. C. Stoney, *Proc. R. Soc. London, Ser. A* **82**, 172 (1909).
12. R. Messier, A. P. Giri, and R. A. Roy, *J. Vac. Sci. Technol. A* **2**, 500 (1984).
13. G.C.A.M. Janssen and J.-D. Kamminga, *Appl. Phys. Lett.* **85**, 3086 (2004).
14. T. Karabacak, A. Mallikarjunan, J. P. Singh, D. Ye, G.-C. Wang, and T.-M. Lu, *Appl. Phys. Lett.* **83**, 3096 (2003).
15. D. Deniz, T. Karabacak, and J. M. E. Harper, *J. Appl. Phys.* **103**, 083553 (2008).
16. I. C. Noyan, T. M. Shaw, and C. C. Goldsmith, *J. Appl. Phys.* **82**, 4300 (1997).

Mater. Res. Soc. Symp. Proc. Vol. 1224 © 2010 Materials Research Society   1224-FF10-22

High-Throughput Optimization of Adhesion in Multilayers by Superlayer Gradient

S.Yu. Grachev, C. Cuminatto, E. Søndergård, E. Barthel
Surface du Verre et Interfaces CNRS/Saint-Gobain UMR 125, 39 quai Lucien Lefranc, 93303
Aubervilliers, Cedex, France

ABSTRACT

We used thickness gradients for high throughput optimization of adhesion in film stacks. The idea is based on the so-called superlayer test where a top layer under high compression exerts a load onto the lower interfaces and may cause delamination and buckling. Thus, on one hand, the thickness gradient of the superlayer results in the *gradient of the load*. On the other hand, the *adhesion gradient* can be realized by changing the thickness of an adhesion enhancer (or an adhesion reducer). When applied in two perpendicular directions (cross-gradient), the gradient of the superlayer in one direction and of the adhesion enhancer in the other, the plane of the sample represents a map where the line of delamination relates the interfacial toughness to the thickness of the enhancer.

In our tests we used Mo superlayers under compressive stress of the order of ~1.5 GPa on a Si wafer with a native oxide. The adhesion reduction was observed with this methodology when Ag layer up to 10 nm thick was deposited onto the substrate prior to Mo deposition. The delamination occurred at Ag thicknesses starting from ~6 nm. This thickness of Ag corresponds to the islands coalescence and formation of a continuous film which immediately results in adhesion reduction. The other test was performed with a step gradient of Ti enhancer placed under a 10 nm thick Ag layer in otherwise the same arrangement. A single test showed that 2.8 Å of Ti was sufficient to improve the adhesion between Ag and SiO_x by several times.

INTRODUCTION

Film stacks are widely used in industrial applications such as optical filters, thermal insulators, hard protective coatings, and in microelectronics. One of the requirements for applied use of thin films is their mechanical stability. However, weak interfacial toughness is often responsible for failures of products. In this context, a reliable and easily applicable test of adhesion is a highly wanted tool for mechanical optimization studies.

The superlayer adhesion test has been used both in adhesion estimation and in fundamental research [1-5]. In this test a superlayer under high compression is deposited directly onto a stack. In this case, there is a tendency for the superlayer to relax the internal stress by breaking a weak interface of the stack and buckling. The load on the interface increases with increase of the elastic energy stored in the stressed layer. Since the elastic energy is proportional to the layer thickness ($E_{elastic} = \sigma^2 h E/(1-\nu)$; σ is the stress, h is the thickness, E is the Young modulus and ν is the Poisson's ratio), a gradient of the thickness of the stressed layer provides a gradient of load. When the thickness of the superlayer is large enough, the delamination may occur.

The use of gradients of the superlayer can give a quantitative value of the threshold thickness for delamination. In this way, the adhesion of interfaces can be compared, although the absolute value of the interfacial toughness is difficult to obtain [6]. Using the same gradient method, a gradient of *adhesion* can also be obtained. Indeed adhesion may depend upon film thickness because an interlayer can either promote or reduce adhesion. Here we demonstrate that cross gradient samples can be obtained where a thickness gradients of the superlayer and an adhesion gradient are applied in orthogonal directions simultaneously in one sample. We show that cross-gradients provide a fast test of the impact of an interlayer thickness on adhesion,

EXPERIMENT

2 inch (50 mm) wafers of Si(100) with a native oxide were used as substrates. All the films were obtained by dc sputter-deposition technique at room temperature. A high vacuum deposition chamber with a base pressure of 1×10^{-5} Pa was used. We applied Mo films under ~1.5 GPa compressive stress as superlayers. For nominally homogeneous Mo layers the thickness varied by less than 10~% across the sample, which insured the control over thickness of the films. More details on experimental conditions are given elsewhere [6].

Thickness gradients were obtained by gradually shading the substrate by a metallic shutter during deposition. The plate was moved either continuously with a constant speed or by discrete steps resulting in the continuous gradient or in a series of stripes of homogeneous thickness, respectively. After rotating the sample by 90° around its normal another gradient was applied (cross-gradient). We realized a gradient of an adhesion modifier (Ti or Ag) in one direction and a gradient of the Mo superlayer in the perpendicular direction.

RESULTS AND DISCUSSION

Figure 1 shows a photo of a sample with a continuous cross-gradient of Ag vs. Mo in the stack Mo/Ag/Si(100). The thickness of Ag is constant along vertical lines and continuously increases from 0 at the lefthand side to 10 nm at the righthand side of the wafer. The thickness of the Mo superlayer increases from 150 nm at the bottow, below the horizontal line, to 450 nm at the upper end of the wafer. The adhesion between Ag and oxides is known to be weak [7]. This is why the addition of Ag reduces the mechanical stability of the stack. At a certain threshold of ~ 6 nm of Ag (the right vertical line) the adhesion became weak enough and delamination and buckling of the layer has occurred. The delamination started in the area with thicker Mo and thicker Ag layers and propagated towards smaller thickness areas before stopping at some threshold thickness which is found to depend upon Ag thickness. Therefore at each thickness of Ag, a threshold thickness of the superlayer can be determined, which reflects the strength of the weak interface or the interfacial toughness. Remarkably, the toughness of the Ag/SiO_x interface decreased rapidly as the thickness of Ag reached the critical value of ~ 6 nm. As the thickness of Ag increased further, the threshold thickness of the superlayer for delamination decreased to ~200 nm. This must be connected to the fact that at room temperature Ag first forms islands which coalesce and form a continuous film at about 5-8 nm of Ag [8, 9]. After coalescence a continuous Ag film forms resulting in the loss of direct Mo/SiO_x coupling and lower adhesion.

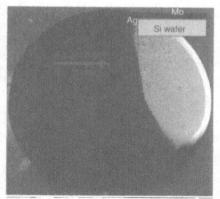

Figure 1. The delamination pattern due to a continuous cross-gradient Ag vs. Mo in a stack Mo/Ag/Si(100). The Ag thickness was changed from 0 at the left red line to 10 nm at the right end of the sample. The thickness of Mo superlayer was changed from 150 nm at the blue line and below to 450 nm at the upper end of the sample. The buckling of the sample was observed in the area with thicker Mo and thicker Ag layer.

Although the adhesion between Ag and SiO_x is weak, it can be improved by adding to this interface an adhesion enhancer, for example, Ti. The effect of Ti is demonstrated in our second experiment. We have prepared a cross-gradient Ti vs. Mo in a stack Mo/Ag(10 nm)/Ti/Si(100). In this experiment, the thicknesses were changed in steps. The thickness of Ti was 0, 1.4, 2.8 and 4.2 Å. The thickness of Mo was chosen to be 206, 412, 618 and 824 nm. The delamination pattern is shown in figure 2.

When no Ti was present at the Ag/SiO_x interface the application of Mo superlayer resulted in delamination all through the Mo thickness. This means that the threshold thickness for delamination was below 206 nm. As 1.4 Å of Ti was added to the weak interface, the threshold thickness had increased to 412 nm. Deposition of 2.8 Å of Ti enhanced the adhesion dramatically and the threshold thickness for delamination elevated to ~824 nm. The adhesion with thicker Ti enhancer was out of range of the superlayer test.

Note that the blisters in both these experiments were of the telephone-cord shape [5]. This shape allows for relaxation of the biaxial stress in both in-plane directions. However the mechanics of the telephone-cord blisters is not well understood and is currently under investigation [5]. In addition to the adhesion optimization, the present tests give access to the morphologies of the telephone-cords for a wide range of thicknesses and adhesions.

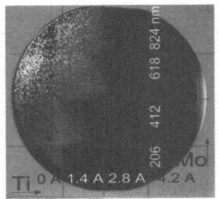

Figure 2. The delamination pattern due to the step cross-gradient Ti vs. Mo in a stack Mo/Ag (10 nm)/Ti/Si(100). The thicknesses are given in the picture. Addition of Ti improved adhesion between Ag and Si(100) systematically and the threshold for delamination increased with Ti thickness.

CONCLUSIONS

The high throughput approach to the adhesion optimization based on the superlayer test of adhesion has been developed. We used a thickness gradient of Mo superlayer in one direction and a thickness gradient of the adhesion modifier in the perpendicular one. The first gradient provided a range of load onto tested interface and the second one resulted in the gradient of interfacial toughness. This geometry allowed for comparative studies of adhesion versus the thickness of an adhesion modifier.

With this test we found that the adhesion between an Ag layer and the native silicon oxide on a Si wafer dramatically decreases at ~ 6 nm of Ag. At this thickness the Ag islands coalesce and start forming a continuous film. The test was also used with the gradient of Ti interlayer between Ag and SiOx. The increase of thickness of Ti resulted in the increased adhesion in this system. A change of Ti thickness with a step of 1.4 Å immediately resulted in the interfacial toughness, which was detected by the superlayer test.

ACKNOWLEDGMENTS

This project is partly funded by the ANR project MatetPro07_247145.

REFERENCES

1. Bagchi, G.Lucas, Z.Suo, and A.Evans, J. Mater. Res. 9 (1994) 1734.
2. M.D.Kriese, W.W.Gerberich, and N.R.Moody, J. Mater. Res. 14 (1999) 3007 and 3019.
3. A.Lee, C.S.Litteken, R.H.Dauskardt, and W.D.Nix, Acta Mater. 53 (2005) 609.
4. M.J.Cordill, D.F.Bahr, N.R.Moody, and W.W.Gerberich, Mater. Sci. Eng. A 443 (2007) 150.

5. C.Coupeau, Mater. Sci. Eng. A 483-484 (2008) 617.
6. S.Yu. Grachev, A. Mehlich, J.-D. Kamminga, E. Barthel, E. Søndergård, Thin Solid Films, submitted.
7. E. Barthel, O Kerjan, P. Nael and N. Nadaud, Thin Solid Films, 473 (2005) 272-7..
8. R. Lazzari, J. Jupille, Surf. Sci. 482-485 (2001), 823.
9. S. Banerjee, S. Kundu, Surf. Sci. 537 (2003), 153.

Mater. Res. Soc. Symp. Proc. Vol. 1224 © 2010 Materials Research Society   1224-GG03-08

Development of a biaxial tensile module at synchrotron beamline for the study of mechanical properties of nanostructured films

E. Le Bourhis[1], B. Girault[1], P.-O. Renault[1], P. Goudeau[1], G. Geandier[2], D. Thiaudière[2], R.N. Randriamazaoro[3], R. Chiron[3], D. Faurie[3], O. Castelnau[3]

[1]LPM-PhyMat, UMR 6630 CNRS, Université de Poitiers, B. P. 30179, 86962 Futuroscope, France
[2]Synchrotron SOLEIL, L'Orme des Merisiers, B. P. 48, 91192 Gif sur Yvette, France
[3]LPMTM, UPR 9001 CNRS, Université Paris-Nord, 93430 Villetaneuse, France

ABSTRACT

We have developed on the DIFFABS-SOLEIL beamline a biaxial tensile machine with synchrotron standard for in-situ diffraction characterization of thin polycrystalline metallic film mechanical response. The machine has been designed to test cruciform substrates coated by the studied film under controlled applied strain field. Technological challenges comprise the fixation of the substrate, the generation of a uniform strain field in the studied (central) volume, the operations from the beamline pilot. Tests on W and W/Cu multilayers films deposited on polyimide substrates are presented.

INTRODUCTION

Understanding the mechanical behavior of nanostructured thin films in relation to their microstructure, in particular to the grain size, is of utmost importance for the development of technological applications [1]. Synchrotron X-ray sources allow for characterizing small volumes of material in an acceptable time schedule. Hence, supported thin films mechanical response has been characterized experimentally in situ by synchrotron X-ray diffraction [2,3] while having in view their microstructure. This is particularly important when the films are multiphased and/or formed by anisotropic elastic crystallites. In such case, when the crystallites are not randomly distributed (for instance in the presence of texture) the film response is expected to be elastically anisotropic. The tests are generally carried out under uniaxial loading, the films being stressed biaxially because of the Poisson ratios mismatch (between the film and substrate). The transversal component of stress is then imposed by the deposited film properties. So far, it is of utmost importance to control both stress components. In that objective we have developed a biaxial loading machine at synchrotron standard allowing for applying in plane normal forces to coated substrates. Experiments on the elastic deformation of isotropic W and on the confined plasticity and fracture of W/Cu multilayers are presented.

BIAXIAL TENSILE MACHINE DESIGN

The tensile machine has been designed to allow for loading along two normal axis cruciform substrates coated by the studied films. The device is to be operated on a beamline at SOLEIL standard (French synchrotron at Gif sur Yvette, France). A micrograph taken at

DIFFABS beamline and a schematic representation of the machine are shown in Figure 1. The machine is compact 3.5 kg in weight, 19x19x8.5 cm³ in size with an empty centre and is designed to allow X-ray diffraction under small angles (no shadowing edges). Two couples of motors and force sensors are fixed to the device frame. The 4 motors can be actuated separately in order to keep the studied area at a fixed position in the goniometer (same studied volume, Figure 1). All force sensors were calibrated using dead weights. The cruciform substrates were coated at their centre only and gripped by a cam rotating in a cylinder fixation. A polymeric substrate is chosen to minimize its mechanical contribution as compared to that of the films. Here, we used 125 µm thick polyimide (sofimide ® from Micel and Kapton ® from Du Pont de Nemours).

 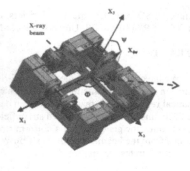

Figure 1 Biaxial tensile device set on DIFFABS-SOLEIL beamline (left) and schematic representation of the biaxial machine, showing the cruciform substrate, gripped by 4 cylinders connected to force sensors and rotated by step motors (right).

One of the most challenging aspects in biaxial testings is the specimen design [4,5]. To perform a biaxial test on sheet material, a cross-shaped specimen is typically used, i.e. cruciform specimen. The objective of the present study was to check that we could perform in situ x-ray strain measurements with high accuracy. Hence, the specimen design was governed by the following constraint: the centre of the substrate specimen must exhibit a homogeneous strain area larger than the irradiated area i.e. a few mm². Using finite elements analysis (CASTEM code from CEA), we checked that under a 100 N equibiaxial loading, a uniform strain (deviation less than 1%) was generated in a central area of 6 mm in radius of the cruciform substrate with 20 mm in width branches and 5 mm toe weld (see Fig. 2) while the used X-ray beam section at sample surface was about 1 x 0.3 mm². It is to be noted that this calculation could be obtained with a 2 dimensional approach in the case of a bare substrate. The actual tested specimens are more complex being coated and require a 3 dimensional approach. In the next section, the measurements are carried out for a coated substrate in order to extract thin film elastic strains and check the biaxiality and uniformity of the loading.

An optical microscope is fixed underneath the machine and allows capturing the bottom surface of the specimens at each load step. Digital image correlation was then used to extract the macroscopic strain of the substrate and to check the uniformity of the strain field. In the future, we could pilot the machine in optical macroscopic central strain. Once the coated substrate is set and gripped in the machine, the test is carried out and controlled in deformation imposed by the motor by incremental steps. The forces are measured at each branch. All electronic signals have been set to SOLEIL standards in order to facilitate the experiments.

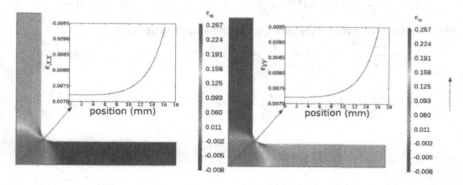

Figure 2 Finite element analysis (FEA) of the strain field in a cruciform substrate loaded under equibiaxial forces of 100 N, ε_{xx} (left) and ε_{yy} (right) components. Insets show the respective strain profile along the line shown on the cruciform.

EXPERIMENTAL RESULTS AND DISCUSSION

<u>Elastic deformation of isotropic W</u>

For the preliminary tests we used W that is elastically isotropic. So far, the Poisson ratio of W is 0.26 (0.284 for the bulk [6]) while we measured 0.37 for the polyimide substrate. These values correspond to a relative Poisson ratio mismatch of 30 and 23% respectively. As discussed above, an uniaxial loading of coated specimens yield both axial and transversal stress components. Instead, the machine was used to impose an equibiaxial stress field to the coated substrate, except for T6 loading state (Fig. 3). T1 to T5 correspond to equibiaxial forces of 10, 37, 63, 81 and 42 N respectively while T6 corresponds to non equibiaxial forces of 72 and 48 N along $\Phi = 0$ and 90 ° respectively. The W thin films were produced at room temperature by physical vapor deposition (PVD) with an Ar+-ion-gun sputtering beam at 1.2 keV (Kaufman ion source) in a NORDIKO-3000 system. The base pressure of the deposition chamber was 7×10^{-5} Pa while the working pressure during films growth was approximately 10^{-2} Pa (For more details see Ref. [3]).

In the case of isotropic materials such as W, the length-scale change is straightforward since the film deforms homogeneously. Measured strains reduce to [7,8]

$$\varepsilon_{\Phi\Psi}^{33} = \left(\frac{1+\nu_f}{E_f}\right) \cdot \left(\sigma_{11}^f \cos^2\Phi + \sigma_{22}^f \sin^2\Phi\right) \cdot \sin^2\Psi - \frac{\nu_f}{E_f} \cdot \left(\sigma_{11}^f + \sigma_{22}^f\right) \quad (1a)$$

E_f and v_f are the film Young modulus and Poisson ratio respectively, σ_{11}^f and σ_{22}^f the principal stress components. Φ is the rotation angle around surface normal and Ψ the angle between the normal of the diffracting planes and the normal of the specimen surface (Fig. 1). For an equibiaxial stress field $\sigma_{11}^f = \sigma_{22}^f = \sigma^f$, strain is independent of Φ and writes

$$\varepsilon_{\Phi\Psi}^{33} = \left(\frac{1+v_f}{E_f}\right)\cdot\left(\sigma^f\right)\cdot\sin^2\Psi - 2\frac{v_f}{E_f}\cdot\left(\sigma^f\right) \tag{1b}$$

We further used two particular angles $\Phi = 0$ and $90°$ for which strains write

$$\varepsilon_{0\Psi}^{33} = \left(\frac{1+v_f}{E_f}\right)\cdot\left(\sigma_{11}^f\right)\cdot\sin^2\Psi - \frac{v_f}{E_f}\cdot\left(\sigma_{11}^f + \sigma_{22}^f\right) \tag{1c}$$

and

$$\varepsilon_{90\Psi}^{33} = \left(\frac{1+v_f}{E_f}\right)\cdot\left(\sigma_{22}^f\right)\cdot\sin^2\Psi - \frac{v_f}{E_f}\cdot\left(\sigma_{11}^f + \sigma_{22}^f\right) \tag{1d}$$

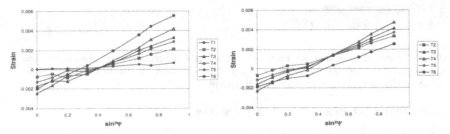

Figure 3 (110)W elastic X-ray strains as a function of $\sin^2\Psi$ for $\Phi = 0°$ ($\varepsilon_{0\Psi}^{33}$ left) and $\Phi = 90°$ ($\varepsilon_{90\Psi}^{33}$ right). 6 different loading states TX were used (T1 to T5 being equibiaxial and T6 being non-equibiaxial, see text).

The linear relationship between strain and $\sin^2\psi$ (so called ε-$\sin^2\psi$ method) is hence established for a single phase elastically isotropic material. So far, most materials show anisotropic elasticity. The mechanical behavior of a polycrystalline thin film is then determined by the distribution of the crystallite orientations within the thin film and the grain interaction [3,9,10]. Figure 3 shows the X-ray strains extracted along both axis ($\Phi = 0$ and $90°$) for 6 loading states (T1 to T6) measuring the Bragg peak shifts for different inclinations of the specimen (different angles Ψ). The strain is calculated using the unloaded state T0 as a reference state: $\varepsilon_{0\psi}^{33} = \ln\left(\sin\theta_{0\psi}^{T0}/\sin\theta_{0\psi}^{TX}\right)$ where $\theta_{0\psi}^{T0}$ is the angular position of the considered diffraction peak for the unloaded state and $\theta_{0\psi}^{TX}$ the corresponding angle for the loaded state TX. As commonly adopted strain is plotted as a function of $\sin^2\Psi$. All curves are linear as expected for an isotropic material [3]. As the load increases the slope of the related curves increases with a loading sequence increasing from T1 to T4 decreasing to T5 while T6 loading state is not equibiaxial. For T2 to T5 loading states, X-ray strains along both directions ($\Phi = 0$ and $90°$) are determined in a difference of less than 20%. Instead for T6 loading state the difference in strain approaches 200%. For equibiaxial stresses Eqs. 1 yield a value of zero strain at $\sin^2\psi = 2v_f/(1+v_f) \sim 0.41$ and 0.44 for v_f of 0.26 and 0.284 respectively in good agreement with the obtained experimental values (0.42 and 0.46 for $\Phi = 0$ and 90° respectively).

Confined plasticity and fracture of W/Cu multilayers

We further investigated the co-deformation of W and Cu sublayers under uniaxial loading. As discussed above a transversal component of stress exists and results from Poisson ratios mismatch. The films were obtained by the same PVD technique, W and Cu sublayer thickness being 6 and 18 nm respectively and with (110) and (111) fiber texture respectively. We took advantage of XRD that is phase selective so that both W and Cu elastic strains can be monitored (Fig. 4). We observe that Cu yields before W under an applied force of about 5 N (estimated stress in Cu sublayer is about 120 MPa, using a simple rule of mixture). TEM micrographs (not shown here) indeed reveal dislocations in Cu sublayer at this step while W sublayers remain free of dislocations (no dislocation detected).

On increasing the applied force to 8 N, W sublayers yield, plastic deformation continuing until micro-fracture happens in W sublayers (strain rate being extremely low and about 10^{-8} s^{-1}). After increasing further the force, macro-fracture runs through the complete multilayer. The TEM cross-sectional view shown in Fig. 4 was obtained at this step and reveals that fracture happens in mode II (slide mode [11]). From the above observations, we suggest that the confined plasticity regime in Cu leads to the accumulation of dislocations at Cu/W interfaces. Next, W sublayers deform with Cu sublayers in planes where Schmid factor maximizes (that is at ~ 45° from load axis) when fracture of the films happens in mode II.

It is worth noting that the overall process yields compressive stresses in both W and Cu sublayers. This final stress state results from the difference between the substrate and film elastic limit. Indeed, the polyimide substrate remains elastic during the test and unloads completely compressing the W and Cu sublayers that plastically deformed.

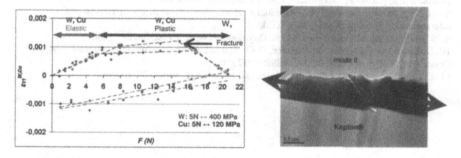

Figure 4 W and Cu X-ray elastic strains as a function of Force F (left) and TEM cross-sectional view after fracture (right).

CONCLUSIONS

We have developed on the DIFFABS-SOLEIL beamline a biaxial tensile machine with synchrotron standard for in-situ diffraction characterization of thin polycrystalline metallic film mechanical response. Preliminary tests on W films deposited on polyimide cruciform substrates show that the device allows for setting a uniform strain field in the characterized volume of a film having a Poisson ratio mismatch with respect to the supporting substrate. Such instrument combined with XRD allows for studying the co-deformation of nanostructured materials like W/Cu multilayers. Further developments are aimed at developing dynamic testing controlled by a ramp of optical strain (determined at substrate backside surface).

ACKNOWLEDGMENTS

Part of this work has been developed in the ANR project entitled Cmonano (ANR-05-NANO-069-03).

REFERENCES

1. M.A. Meyers, A. Mishra, D.J. Benson, Prog. Mater. Sci. 51 (2006) 427.
2. J. Böhm, P. Gruber, R. Spolenak, A. Stierle, A. Wanner, E. Arzt, Rev. Sci. Inst. 75 (2004) 1110.
3. D. Faurie, P.-O. Renault, E. Le Bourhis, P. Villain, Ph. Goudeau, and K. F. Badawi, Thin Solid Films, 469-470 (2004) 201.
4. A. Hannon, and P. Tiernan, J. Mater. Proc. Techn. **198**, 1 (2008).
5. S. Demmerle, J.P. Boehler, J. Mech. Phys. Solids **41**, 143 (1993)
6. P.O. Renault, K.F. Badawi, L. Bimbault, P. Goudeau, E. Elkaïm, J.P. Lauriat, Appl. Phys. Lett. 73 (1998) 1953
7. V. Hauk (1997), Structural and residual stress analysis by non destructive methods: Evaluation, application, assessment. Elsevier Science, Amsterdam.
8. I.C. Noyan, J.B. Cohen (1987) Residual stresses. Measurements by diffraction and interpretation, Springer Verlag, New York.
9. S. Matthies, H. G. Priesmeyer, M. R. Daymond, J. Appl. Cryst. **34**, 585 (2001).
10. D. Faurie, O. Castelnau, R. Brenner, P.-O. Renault, E. Le Bourhis, Ph. Goudeau, J. Appl. Cryst. **42** (2009) doi:10.1107/S0021889809037376.
11. E. Le Bourhis (2008), Glass mechanics and technology, Wiley, Weinheim.

Mater. Res. Soc. Symp. Proc. Vol. 1224 © 2010 Materials Research Society   1224-FF05-40

Micromechanical Testing of Nanostructured NbTiNi Hydrogen Permeation Membranes

Tetsuya Kusuno[1], Yusuke Shimada[1], Mitsuhiro Matsuda[1], Masaaki Otsu[1], Kazuki Takashima[1], Minoru Nishida[2], Kazuhiro Ishikawa[3], Kiyoshi Aoki[3]

[1]Department of Materials Science and Engineering, Kumamoto University, 2-39-1, Kurokami, Kumamoto, Japan
[2]Department of Applied Science for Electronics and Materials, Kyushu University, Fukuoka, Japan
[3]Department of Materials Science and Engineering, Kitami Institute of Technology, Hokkaido, Japan

ABSTRACT

Nb-Ti-Ni alloy is one of the candidates for hydrogen permeation membranes. The hydrogen permeability of a membrane depends on its thickness, and mechanical properties such as the fracture toughness of the membrane are important to ensure reliability and durability. In the present work, micro-mechanical tests have been carried out for melt-spun Nb-Ti-Ni thin films consisting of amorphous and nano-crystalline phases. The relationship between the mechanical properties of the melt-spun films and the microstructural changes occurring in the films due to heat treatment has been also discussed. The Nb-Ti-Ni alloy thin films were prepared by the melt-spun technique and then heat-treated at 873-1173 K. Micro-sized cantilever specimens with dimensions of $10 \times 10 \times 50$ μm^3 were prepared by focused ion beam (FIB) machining. Fracture tests were carried out using a mechanical testing machine for the micro-sized specimens; the testing machine was developed by us. In addition, microstructures were observed by transmission electron microscopy (TEM). The fracture toughness (K_Q) value decreased up to 823 K, and it increased above 1173 K. The specimen heat-treated above 1173 K showed ductile fracture. The fracture morphology of the specimen heat-treated up to 1023 K showed grain boundary fracture characteristics, and that of the specimen heat-treated at 1173 K changed to transgranular fracture.

INTRODUCTION

Recently, hydrogen energy has been attracted a great deal of attention as clean energy instead of fossil fuel. The fuel cell utilizing the hydrogen gas has been extensively investigated and developed. However, there are many problems, such as the damage of the Pt electrode in the cell with impurities in the hydrogen gas. Therefore the separation and purification of the hydrogen gas with hydrogen permeation membrane alloys are indispensable. Pd-Ag alloys have been commercially used as hydrogen permeation membranes to separate and purify hydrogen gas. Since Pd is quite expensive and its resources are scarce, Hashi et al. [1-3] recently developed Nb-Ti-Ni alloy consisting of bcc-Nb and B2-TiNi eutectic microstructures with a good combination of high hydrogen permeability and excellent hydrogen embrittlement resistance as non-Pd-based hydrogen permeation alloys with low cost and high performance. The high hydrogen permeability is due to the bcc-Nb solid solution phase, while the excellent hydrogen embrittlement resistance is attributable to the B2-TiNi phase. The hydrogen permeability of a membrane increases with a decrease in its thickness [1, 4]. Therefore, the mechanical properties of permeation membranes, such as fracture toughness, are important to ensure reliability and durability. Liquid quenching is suitable for the fabrication of thin-films, because it is cost-effective and easily applicable to mass-production. Hydrogen permeability increases at high temperature. Moreover, it is necessary to consider the microstructural changes occurring in a membrane at high temperature as the microstructure of the Nb-Ti-Ni alloy also affects its mechanical properties. Therefore, the relationship between the microstructural changes by heat treatment and the mechanical properties of this alloy is very important. In the present work, micromechanical tests have been conducted for melt-spun $Nb_{40}Ti_{30}Ni_{30}$ ribbons, and the relationship between the mechanical properties and microstructural changes by heat treatment has been investigated.

EXPERIMENT

In this study, the $Nb_{40}Ti_{30}Ni_{30}$ ribbons were prepared by liquid quenching. Differential scanning calorimetry (DSC) was performed to examine the crystallization process. Subsequently, the $Nb_{40}Ti_{30}Ni_{30}$ liquid-quenched ribbons were heat-treated between 823 and 1173 K on the basis of the results of DSC. Micro-sized cantilever specimens were cut from the ribbons by focus ion beam (FIB) machining. Figure 1 shows a scanning electron micrograph of the micro-sized specimen. The length (L), breadth (B) and width (W) of specimen were 50, 10 and ~20 μm, respectively. Furthermore, a notch with a width of 0.5 μm and a depth of 5 μm was also introduced into the micro-sized specimens by FIB. The notch position was set at 10 μm from the fixed end of the specimens. The loading position was located 30 μm from the notch. Fracture tests were conducted at room temperature in air using a mechanical testing machine [5] for the micro-sized specimens. The load resolution of this testing machine was 20 μN, and the displacement resolution was 10 nm. The loading position could be adjusted using a precise X-Y stage at a translational resolution of 0.05 μm. The fractured surface and the sides of the specimens were observed after micro-mechanical testing by scanning electron microscopy (SEM). Transmission electron microscopy (TEM) observations were also performed to investigate the microstructural changes by heat treatment.

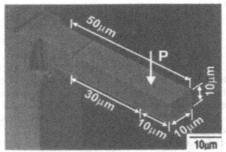

Figure 1. Scanning electron micrograph of micro-sized cantilever beam specimen.

RESULTS AND DISCUSSION

Figure 2 shows load-displacement curves of the micro-sized cantilever specimens, measured during the fracture toughness tests. Figure 3 shows the fracture toughness (K_Q) values calculated from the fracture test results. In this study, the fracture toughness tests were not performed following the ASTM standard, so the fracture toughness values are denoted by K_Q. The K_Q values were calculated using the following equations, where K is the stress intensity for a notched cantilever beam.

$$K = \frac{6PS}{W^2B}\sqrt{\pi a}F(a/W), \quad (a/W < 0.6)$$ (1)

where

$$F(a/W) = 1.122 - 1.40 - 1.40(a/W) + 7.33(a/W)^2 - 13.08(a/W)^3 + 14.0(a/W)^4$$ (2)

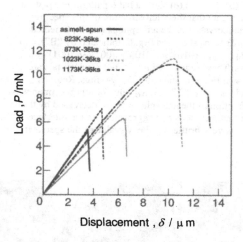

Figure 2. Load-displacement curves measured during fracture testing.

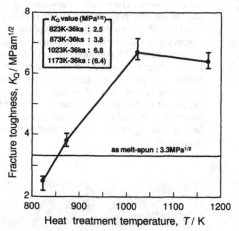

Figure 3. Fracture toughness value (K_Q) measured during fracture toughness testing.

The K_Q value of the as melt-spun specimen was 3.3 MPa$^{1/2}$. Brittle fracture was observed in the specimen. The K_Q value decreased by heat-treatment up to 823 K, and then it increased above 873 K. Ductile fracture was observed in the specimen annealed at 1173 K.

Figure 4 shows a scanning electron micrograph of the fractured surface of each specimen. Three types of fractured surfaces were observed. A flat fractured surface, i.e., a typical brittle fracture surface was observed for the as melt-spun ribbon and the specimen heat-treated at 823 K (Fig. 4(a)). Many small asperities were observed on the fractured surface of the specimens heat-treated at 873 K and 1023 K (Fig. 4(b)). For the specimen heat-treated at 1173 K (Fig. 4(c)), the size of the asperities on its fractured surface increased.

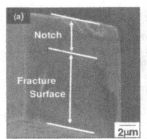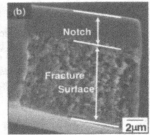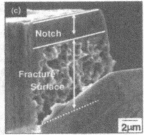

Figure 4. Scanning electron micrographs of fractured surface of specimens (a) as melt-spun, (b) heat-treated at 1023 K, and (c) heat-treated at 1173 K.

Figure 5 shows the bright field (BF) image and the diffraction patterns of the as melt-spun specimen. The patterns shown in Figs. 5 (b), (c), and (e) show a hallo pattern, bcc pattern, and Debye ring, respectively. These patterns indicate that the microstructure of this specimen consisted of a crystallized area (Nb phase or nanocrystal) in the amorphous matrix. Oriented nanocrystals were formed in the specimen at 823 K, as shown in Fig. 6 (indicated by "B"), and the cube-cube (C-C) relationship of $[100]_{B2}$ // $[100]_{bcc}$, $(010)_{B2}$// $(010)_{bcc}$, $(001)_{B2}$// $(001)_{bcc}$ was found between TiNi and Nb(Ti) solid solutions. These regions with oriented nanocrystals are called aggregates. The diffraction pattern show in Fig. 6(c) indicates Debye rings. Regions consisting of non-oriented nanocrystals are indicated by "C." The K_Q value of this specimen was the lowest in this study, because the volume fraction of the amorphous phase decreased by the heat treatment, and resulted in the formation of nanocrystals and aggregates. As the interface between the aggregates and the nanocrystals was very brittle [6], the K_Q value of the specimen decreased.

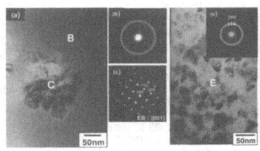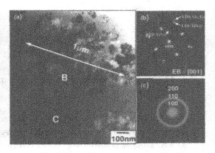

Figure 5. (a) and (d) Bright field images and corresponding electron diffraction patterns of (b) area B and(c) area C, and (e) area E of as melt-spun $Nb_{40}Ti_{30}Ni_{30}$ ribbon.

Figure 6. (a) Bright field image and corresponding electron diffraction patterns of (b) area B, and (c) area C of specimen heat-treated at 823 K.

Figure 7 shows TEM micrographs of the specimens heat-treated at 873-1023 K. The corresponding electron diffraction pattern shown in Fig. 7(b) indicates that the nanocrystal area disappeared completely, and crystalline grain growth was observed in the aggregate in keeping with the C-C relationship. The K_Q value of this specimen was found to increase because of the formation of the aggregates and the disappearance of the brittle interface between the aggregates and the nanocrystals. The size of the asperities and the grain size of the fractured surface of the specimen heat-treated at 1073 K was about 150 nm. This result suggests that the crack propagated along the grain boundary in the aggregates. Since the grain structure changed with increasing heat treatment temperature and the K_Q value increased, the crack growth path was complicated in the specimen heat treatment temperature between 873 K and 1023 K.

Figure 8 shows the bright field (BF) image and the diffraction patterns of the specimen heat-treated at 1173K. The pattern shown in Fig. 8 (b) show Nb phase. There was no C-C relationship between the TiNi and Nb(Ti) solid solutions, and only crystalline grain growth was observed, the grain size being 200-400 nm. The size of the asperities on the fractured surface shown in Fig. 4(c) was larger than the grain size. This suggests that the crack growth path changed as the heat treatment temperature increased; thus, the fracture morphology changed from grain boundary fracture to transcrystalline fracture. It is possible that this change can be controlled to improve the ductility of the specimen.

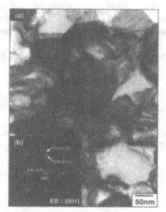

Figure 7. (a) Bright field image and (b) corresponding electron diffraction pattern of specimen heat-treated at 1023 K.

Figure 8. (a) Bright field image and corresponding electron diffraction pattern of (b) area B of specimen heat-treated at 1173 K.

CONCLUSIONS

Micromechanical tests were conducted on as melt-spun ribbons and heat-treated $Nb_{40}Ti_{30}Ni_{30}$ ribbons, and the relationship between the mechanical properties of the specimens and their microstructural changes was investigated. The microstructure changed progressively with heat treatment. Oriented nanocrystals were formed at 823 K, and the cube-cube relationship of $[100]_{B2}$ // $[100]_{bcc}$, $(010)_{B2}$// $(010)_{bcc}$, $(001)_{B2}$// $(001)_{bcc}$ was found between TiNi and Nb(Ti) solid solutions. The nanocrystals formed aggregates with increasing heat treatment temperature. Fracture toughness (K_Q) values also changed with the evolution of the microstructure. The K_Q value decreased up to 823 K because of the formation of a brittle interface between the aggregates and nanocrystals. The K_Q value increased above 873 K. The crack growth path in the specimen heat-treated between 873 and 1023 K propagated along the grain boundary. Therefore, the K_Q value increased as grain growth occurred with increasing heat treatment temperature. Ductile fracture occurred in the specimen heat-treated at 1173 K. It is considered that the fracture morphology changed from grain boundary fracture to transgranular fracture.

ACKNOWLEDGMENTS

This work was supported by Grants-in-Aid for Scientific Research (B) from the Japan Society for the Promotion of Science (JSPS).

REFERENCES

1. K. Hashi, K. Ishikawa, T. Matsuda, K. Aoki, Journal of Alloys Compounds, **368** (2004), pp. 215-220; **425** (2006), pp. 284-290; Materials. Transactions, **46** (2005), pp. 1026-1031.
2. S. Tokui, K. Ishikawa, K, Aoki, Journal of the Japan Institute of Metals, **71** (2007), pp. 176-180.
3. K. Ishikawa, K. Aoki, Journal of the Japan Institute of Metals, **72**, No.10 (2007), pp. 845-849.
4. K. Kishida, Y. Yamaguchi, K. Tanaka, H. Inui, S. Tokui, K. Ishikawa, K. Aoki, Intermetallics, **16** (2008), pp. 88-95.
5. K. Takashima and Y. Higo, Fatigue and Fracture of Engineering Materials and Structures, **28** (2005), pp. 703-710.
6. Y. Shimada, M. Matsuda, Y. Kawakami, M. Otsu, M. Takashima, M. Nishida, K. Ishikawa, K. Aoki, Journal of the Japan Institute of Metals, **72**, No.12 (2008), pp. 1015-1020.

Mater. Res. Soc. Symp. Proc. Vol. 1224 © 2010 Materials Research Society   1224-FF05-11

Mechanical Properties of Nanostructured hard coating of ZrO$_2$.

R. F. Sabirianov[1], F. Namavar[2], X.C. Zeng[3], J. Bai[3], W.N. Mei[1]

[1]Department of Physics, University of Nebraska - Omaha, Omaha, NE 68182 USA
[2]Department of Orthopaedic Surgery and Rehabilitation,University of Nebraska Medical Center, Omaha, NE 68198 USA
[3]Department of Chemistry and Center for Materials Research Analysis,University of Nebraska - Lincoln, Lincoln, NE 68588 USA

Abstract

Nano-crystalline films of pure cubic ZrO2 have been produced by ion beam assisted deposition (IBAD) processes which combine physical vapor deposition with the concurrent ion beam bombardment in a high vacuum environment and exhibit superior properties and strong adhesion to the substrate. Oxygen and argon gases are used as source materials to generate energetic ions to produce these coatings with differential nanoscale (7 to 70 nm grain size) characteristics that affect the wettability, roughness, mechanical and optical properties of the coating. The nanostructurally stabilized chemically pure cubic phase has been shown to possess hardness as high as 16 GPa and a bulk modulus of 235 GPa. We examine the mechanical properties and the phase stability in zirconia nanoparticles using first principle electronic structure method. The elastic constants of the bulk systems were calculated for monoclinic, tetragonal and cubic phases. We find that calculated bulk modulus of cubic phase (237GPa) agrees well with the measured values, while that of monoclinic (189GPa) or tetragonal (155GPa) are considerably lower. We observe considerable relaxation of lattice in the monoclinic phase near the surface. This effect combined with surface tension and possibly vacancies in nanostructures are sources of stability of cubic zirconia at nanoscale.

Introduction

Tough, wear resistant, refractory zirconia ceramics are used to manufacture parts operating in aggressive environments, like extrusion dyes, valves and port liners for combustion engines, low corrosion, thermal shock resistant refractory liners or valve parts in foundries. High temperature ionic conductivity makes zirconia ceramics suitable as solid electrolytes in fuel cells and in oxygen sensors. Good chemical and dimensional stability, mechanical strength and toughness, coupled with a Young's modulus in the same order of magnitude of stainless steel alloys is the origin of the interest in using zirconia as a ceramic biomaterial.[1]

Zirconia ZrO$_2$ has three distinct polymorphous phases at ambient pressure. It exhibits cubic phase (fluorite $Fm3m$) above 2640K and melts at about 2950K. The cubic zirconia transforms into tetragonal structure ($P4_2/mnc$) below 2640K. The tetragonal zirconia changes into monoclinic structure ($P2_1/C$) at 1440K. Because of its high strength and stability at high temperature ZrO$_2$ has numerous industrial applications such as oxygen sensor, high temperature fuel cell, and gate material for semiconductor device. Among them, cubic and tetragonal polymorphs of zirconia are not stable at room temperature in the bulk samples. The doping of zirconia with trivalent elements is generally used to stabilize these phases. For practical applications it is very important to form dense hard films of zirconia, while most of studies were performed in the powder form. Recently, there were several reports of forming tetragonal [2] and cubic phases [3] in nanoparticles of small size. Among them, the nanostructurally stabilized undoped ZrO$_2$ samples were produced using ion-beam assisted deposition (IBAD) technique which are voids free, high density, hard (16GPa), and have good adhesion to the substrate. Striking findings for ZrO$_2$ films produced by IBAD is the ability to change surface properties in large degree by the deposition conditions: we found that zirconia films exhibit superhydrophilic properties, i.e. the drop of water spreads completely on the surface with zero degree apparent contact angles [4]. The key factor on the increased wetting is a large ratio between the actual and the apparent surface area (Wenzel's ratio). Note, that the surface rms roughness in our samples is very small (2-20nm), i.e. the samples are smooth from the macroscopic point of view. Hence this enhanced "roughness" of the surface renders change the in electric and mechanical properties of zirconia surface. As we know the solid-gas surface tension that affects the wetting properties can be derived from the basic properties of a solid, hence it is possible to deduce by using the information of wetting angles at the solid-liquid interface surface which will are crucial to design the material combining wettability, ductility and hardness.

In particular, ZrO_2 samples fabricated by using IBAD technique offer an opportunity to manipulate the surface morphology and texture of the deposited films [5]. From monitoring the sensitive growth conditions such as temperature, oxygen pressure and substrate, the formation of the pyramidal-like surface structures with the three-fold symmetry has been observed [4], which is probably connected with the over-the-edge diffusive process. We believe this is a manifestation of in-plane rotational disorder in this well-textured film, which is reminiscent to the well-known Stranski-Krastanov mode on the strained surfaces [6,7].

The formation of vacancies, which might be stabilized at the interfaces and grain boundaries, could also be assisted by high-energy ions in the IBAD process. The gradient of stress across the nanostructure might result in the energy profile favorable for keeping the vacancies in the film, which occurs if the excessive strain induced by the vacancy formation on the lattice is negative, or equivalently the stress inside the nanostructure is compressive. EXAFS measurements have been performed on the IBAD samples and confirmed the presence of vacancies [8]. Also, recent findings that nanocrystalline zirconia can be amorphized by ion irradiation suggests that there is a delicate balance between various contribution to the total free energy, namely, bulk, surface/interface, and the defects, governing the stability of these phases. On top of that, recent experimental data on nanostructurally stabilized cubic and tetragonal phase suggest that the dominant mechanism of stability at nanoscale is driven by formation of vacancies [9].

In the present article we present experimental and theoretical studies of nanostructurally stabilized pure cubic zirconia. The paper is organized as follows: First, we present experimental analysis of nanostructured ZrO_2 samples fabricated by using ion beam assisted deposition (IBAD). Second, we compare measured elastic properties of pure bulk zirconia phases with that calculated using ab-initio density functional methods. Finally, we discuss the structural and elastic properties of surfaces of various zirconia polymorphs.

Results and Discussion
Results of experimentations

We have synthesized [4] an ultrahydrophilic (Figure 2) hard pure (without chemical stabilizer) cubic (diamond simulants) zirconia films with a roughness of a few to 20 nm using IBAD. IBAD combines physical vapor deposition with concurrent ion beam bombardment (from an ion gun) in a high vacuum environment to produce films with superior properties. These films are then "stitched" to the orthopaedic artificial implant materials with differential nanoscale (5 to 50 nm grain size) characteristics that affect the wettability [4], roughness, and mechanical properties of the coating. Details of the physical and chemical properties of these nano-engineered coatings are described in references [3] and [4].

The electron diffraction and X-Ray [4] support the formation of cubic phase of ZrO_2. The grain size is ~10 nm based on the dark field images shown in figure 1. The observed grain size is reflected in the small rms roughness derived from the atomic force microscopy (AFM) images shown in figure 2. Figure 2(b) shows that this surface exhibit super-hydrophilic behavior, which is demonstrated by a small contact angle of about 5° for deionized water. The measurements were taken a few days after deposition at room temperature. The contact angle for the conventional zirconia coated hip ball is much larger (shown in figure 2 (a).

The nanoindentation measurements shown in Figure 3 demonstrate a large nanohardness of our ZrO_2 samples. The derived elastic bulk modulus is 239±12GPa.

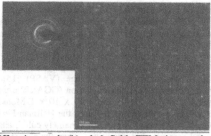

Figure 1. (a) Electron diffraction and (b) dark-field TEM image demonstrating formation of nanocrystalline ZrO_2, with 5–10 nm size, at room temperature.

Figure 2. (a) Contact angle of the water on the ZrO_2 Hip ball (51˚ for water) and (b) on nanostructured cubic zirconia. Contact angle of a 0.25 µL water droplet on structurally stabilized nanocrystalline cubic ZrO2 film produced at room temperature and measured a few days after deposition. A contact angle of 4 ± 0.5° measured after 10 months, keeping the sample in a laboratory environment. (c) AFM of the typical sample. Rms roughness is 2.4nm for this sample.

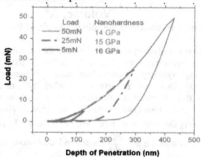

Figure 3.   Nanoindentation measurements of nanoengineered cubic ZrO_2.

Elastic properties of pure phases

The elastic properties of various polymorphs of zirconia attracted significant attention.[10,11,12,13,14] The pure phase of monoclinic zirconia have been studied experimentally, while elastic constants of *pure* crystalline zirconia have not yet been measured directly, thus far, because of the difficulty in producing stabilizer free cubic and tetragonal zirconia. Trivalent-based material, such Y_2O_3 or CaO, with doping ratios up to 20% are used to stabilize these phases. Previously reported elastic constants of cubic and tetragonal zirconia are based on extrapolation to zero concentration of experimentally measured elastic constants of zirconia as function of doping ratio at different temperatures. As a consequence, the elastic constants of stabilizer-free pure zirconia could be quite different [10,11].

We perform first-principle the DFT calculations using the projector augmented-wave (PAW) method implemented in the Vienna *Ab-Initio* Simulation Package (VASP) [15]. The exchange-correlation potential is treated in the generalized gradient approximation (GGA). We use the energy cut-off of 350 eV for the plane wave expansion of the PAWs and a 10 x 10 x 1 Monkhorst-Pack grid for k-point sampling. All the structural relaxations are performed until the Hellman-Feynman forces on the relaxed atoms become less than 10 meV/Å. Three phases of ZrO_2, namely cubic, tetragonal and monoclinic, are relaxed at zero temperature and zero pressure conditions. Then the elastic constants are calculated from the above-obtained optimized structures by setting either the stress or the strain to a finite value, followed by re-optimizing any free parameters and then calculating the strain or stress. At this moment we would like to emphasize, even though there were large amount of studies on ZrO_2, as far as we know the one carried out here is the first systematic work in which relaxed structures of all the phases and their elastic properties are investigated in such details and using unified state-of-the-art approach.

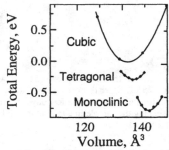

Figure 4. The energy of the system as function of the volume calculated from first principles.

In case of systems with several polymorphs, like zirconia, the structure which has the lowest total energy is the most stable one at zero temperature. Fig. 4 shows that the most stable structure of zirconium dioxide is monoclinic, while high symmetry cubic is least stable. However, the stable polymorph does not necessarily have desired hardness (or, for example, higher bulk modulus) or other mechanical properties. The bulk modulus, B, of a crystal is calculated as a response to a uniform dilation, $\delta=3\Delta a/a$, (energy $U = \frac{1}{2}B\delta^2$).[16] Although the total energy-volume graph (Fig. 4) shows that the cubic phase does not have the lowest energy among the three known phases, but it has the largest curvature meaning that it has the highest elastic bulk modulus ($B = V\dfrac{\partial^2 U}{\partial^2 V}$ is second derivative of energy U with respect to dilation), and indicates that it is hardest among the three phases, while tetragonal is least hard.

Harder materials are more suitable for various applications (bearing coating, implants etc.). Although bulk moduli are indicative of mechanical hardness, the more elaborate analysis requires knowledge of a full set of elastic stiffness constants, reflecting elastic response to deformations along different directions. Cubic phase might have the "stiffest' bonds but has soft modes as well, which results in a phase transition to a structure with lower bulk modulus. Elastic stiffness constant tensor, C, connects stress component, X, with corresponding strain components, ε as following: $\vec{X} = \vec{\vec{C}}\vec{\varepsilon}$, where elastic constant tensor is a 6 by

6 matrix with C_{ij} elements $(i,j=1-6)$. The elastic stiffness constants are connected to bulk modulus and Young's modulus.[17] For example, in the cubic crystal bulk modulus is related to stiffness constants as $B = \frac{1}{3}(C_{11} + 2C_{12})$.[16]

Table 1. Calculated bulk modulus of zirconia.

	Cubic	Tetragonal	Monoclinic
Volume, A³	132.3	137.5	142.9
Bulk modulus, GPa	237	155	189
Expt. Bulk Moduli, GPa	239±12		212

The elastic constants were calculated numerically based on the relaxed structures with setting the strain to a finite value and calculating the resulting stress. By appropriate choose of the applied deformations, the complete set of thirteen elastic constants of the monoclinic zirconia is determined, and the results are presented in Table 2. From our calculation, presented in Table 1 bulk modulus of cubic phase (236.3GPa) is larger than the one of Monoclinic (189.43) and Tetragonal (156.0GPa) phases. Theoretical calculations of elastic constants of cubic and tetragonal zirconia have been carried out previously using first-principles approaches.[12,14] However, to the best of our knowledge, the elastic constants of monoclinic zirconia has not yet been reported. Therefore, the presented first-principles calculation of elastic constants can be of useful to the understanding of differences in mechanical properties among three polymorphous phases of zirconia.

Table 2. Calculated elastic constants of zirconia.

	C_{11}	C_{22}	C_{33}	C_{44}	C_{55}	C_{66}	C_{12}	C_{13}	C_{14}	C_{23}	C_{25}	C_{36}	C_{46}
Monoclinic	342	335	302	80	73	115	142	102	28	120	-4	-1	-14
Cubic	402			177			105						

The calculated elastic constants, bulk modulus shown in Tables 1 and 2, and lattice parameters of three polymorphs of zirconia in the bulk gives understanding of their relative stability and details of elastic properties. They can be compared directly with experimental measurements of bulk samples and in the analysis of elastic properties of nanostructures.

Surface properties of zirconia

The theoretical results presented above are dealing with properties of zirconia in the bulk. The nanostructures of zirconia at sizes of few nanometers have strong effects from the surface. Particle with size of 5nm will have almost half of its atoms in two-three surface layers. Since it is important to take into account surface effects when describing properties of nanostructured films, it is particularly essential for zirconia because three polymorphs can be related to cubic phase with appropriate deformation. The surface stress is one of the major parameters which can be used to understand properties such as adatom mobility, wettability, and pressure imposed inside of nanoparticle, etc. The surface stress can be written as [18]

$$\sigma_{ij}^{surf} = \frac{1}{A}\frac{dE^{surf}}{d\varepsilon_{ij}} = \tau_{ij} + \frac{1}{2}S_{ij\alpha\beta}\varepsilon_{\alpha\beta}$$

where ε_{ij} is the surface-strain tensor and A is the equilibrium surface cell area, τ_{ij} is the intrinsic surface stress tensor and $S_{ij\alpha\beta}$ is the surface excess elastic modulus. In linear elastic approximation the elastic constants are independent of strain (higher order terms in the strains can be neglected) and the surface stress depends linearly on the strain [18]. Based on the above definition and linear approximation we performed calculation of the surface tension calculations for the pure (001) surface. The uniform compressive strain is applied in the plane of the slab to find the elastic response of the system. The slab is allowed to relax in the perpendicular direction. Uniform compressive strain, ε, in the (001) plane of the slab causes the increase in the distance between Zr and O layers along (001) direction perpendicular to the (001) plane (as a solid tends to preserve the volume). For cubic it gives 0.48N/m, while for monoclinic 0.65N/m derived from the Bain path shown in figure 5 [7].

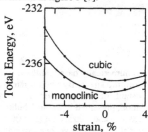

Figure 5. The energy of the slab as a function of the strain.

Conclusions

We have shown that nanocrystalline ZrO$_2$ can be stabilized at nanoscale in cubic phase. We show that cubic phase of zirconia has higher bulk modulus, and particular terminations of cubic zirconia have lower surface energy. Nanocrystalline/nanocolumnar films may have higher surface energy and being more hydrophilic.

References

1. L. Gremillard and J. Chevalier, Key Engineering Materials, 361-363, 791 (2008).
2. S.Shukla, S. Seal, *Int. Mater. Rev. 50*, 45–64 (2005).
3. F. Namavar, G. Wang, C. L. Cheung, R. Sabirianov, X. C. Zeng, W.N. Mei, J. Bai, J. R. Brewer, H. Haider and K.L. Garvin, Nanotechnology 18, 415702 (2007).
4. F. Namavar , C. L. Cheung, R. F. Sabirianov, W.-N. Mei, X. C. Zeng, G. Wang, H. Haider and K. L. Garvin, NanoLetters, 8, 988 (2008).
5. L. Dong and D.J. Srolowitz, J. Appl. Phys., 84, 5261 (1998).
6. C.-H. Chiu, Z. Huang, and C. T. Poh, *Physical Review Letters*, 93 (13), 36105 (2004).
7. R. F. Sabiryanov, M.I. Larsson, K.J. Cho, W.D. Nix, B.M. Clemens, Phys.Rev.B, 67, 125412-5412 (2003).
8. Y. L. Soo, P. J. Chen, S. H. Huang, T. J. Shiu, T. Y. Tsai, Y. H. Chow, Y. C. Lin, S. C. Weng, S. L. Chang, G. Wang, Chin Li Cheung, R. F. Sabirianov, W. N. Mei, F. Namavar, H. Haider, K. L. Garvin, J. F. Lee, H. Y. Lee, and P. P. Chu, J. Appl. Phys. 104, 113535 (2008).
9. S. Fabris, A. T. Paxton and M. W. Finnis, *Acta Mater*. 50, 5171 (2002).
10. H.M. Kandil and J.D Greiner, and J.F. Smith, J. Am. Ceram. Soc. 67, 341 (1984).
11. R.P. Ingel and D. Lewis III, J. Am. Ceram. Soc. 71, 265 (1988).
12. E.V. Stefanovich, A.L. Shlunger, C.R.A. Catlow, Phys. Rev. B 49, 11560 (1994).
13. P. Aldebert and J.P. Traverse, J. Am. Ceram. Soc. 68, 34 (1985).
14. M. Wilson, U. Schonberger and M. Finnis, Phys. Rev. B 54, 9147 (1996).
15. G. Kresse and J. Furthmüller, Phys. Rev. B, 54, 11169 (1996).
16. C. Kittel, *Introduction to Solid State physics*, 3rd ed., John Wiley & Son, (1967).
17. J. Eichler, U. Eisele, and J. Rödel, *J. Am. Ceram. Soc.*, 87 (7), 1401 (2004).
18. R. Shuttleworth, Proc. Phys. Soc. London, Sect. A 63, 444 (1950).

Mater. Res. Soc. Symp. Proc. Vol. 1224 © 2010 Materials Research Society 1224-FF09-07

The Micromechanisms of Deformation in Nanoporous Gold

Rui Dou and Brian Derby
School of Materials, University of Manchester, Grosvenor Street,
Manchester, M1 7HS, UK.

ABSTRACT

We have carried out a TEM investigation of the micromechanisms of deformation in these nanoporous gold specimens after compression testing. We find that the nanoporous specimens show deformation localized to the nodes between the ligaments of the foamed structure, with very high densities of microtwins and Shockley partial dislocations in these regions. These deformation structures are very different from those seen after solid nanowires are tested in compression, which show very low dislocation densities and a few sparsely distributed twins. However, similar dislocation structures to those found in the nanoporous specimens are observed in the larger nanowires when they are deformed in bending. The currently accepted model for the deformation of nanoporous gold, implicitly assumes that the deformation of these structures is by bending near the nodes where ligaments intersect. We hypothesis that the much higher dislocation densities seen in both the nanoporous gold and the nanowires deformed in bending are evidence for the presence of geometrically necessary dislocations in these deformed structures.

INTRODUCTION

In polycrystalline metals there is a well known dependence of yield strength on grain size. Small single crystal metal micropillars or nanowires show a similar but unrelated size effect and at the smallest diameters these are considerably stronger than the corresponding bulk metal. [1-3] Studies of the mechanical properties of nanoporous metallic foams (chiefly gold) also report an increase in yield strength with decreasing ligament diameter with the smallest diameter ligaments having strength in excess of 1 GPa. [4-10] Despite this apparent similarity in the mechanical behavior of metallic micropillars/nanowires and nanoporous gold, the geometry of plastic deformation is different in the two cases. Most of the studies of metal micropillars have reported compression tests of isolated pillars fabricated by focused ion beam machining (FIB). These structures are seen to deform by single or multiple slip, localized to a few slip planes or shear bands, and display characteristic surface steps. Nanoporous gold, however, is believed to deform by mechanisms originally ascribed to the plastic collapse of much larger scale foams, through the formation of plastic hinges at nodes in the structure. [11, 12]

There is currently no universal consensus as to the mechanisms that result in the scale dependent behavior observed with the deformation of micropillars or nanoporous gold. Greer and Nix [13] have proposed a mechanism in which the close proximity of a free surface and the associated image forces remove dislocations from the structure, requiring continuous nucleation of new dislocations to maintain deformation. Recent *in situ* deformation work in the transmission electron microscope (TEM) has observed dislocation generation an annihilation that supports this possible mechanism [14, 15].] In nanoporous gold the accepted structural model for foam collapse [11, 12] requires the presence of strain gradients and in earlier work we have shown that a strain gradient hardening model could explain the high stresses found in nanoporous gold [16]. However, TEM studies of nanoporous gold have not searched for evidence of dislocation storage after the deformation of these materials [17, 18] and thus we are unable to determine whether such a mechanism is possible. Here we present a TEM and HRTEM study of deformed nanoporous gold produced by templated electrodeposition and dealloying. These results are compared with those obtained from deformation experiments on solid gold nanowires with diameters in the range 30 – 70 nm.

EXPERIMENT

Solid gold nanowires were formed by electrodeposition into the pores of ordered anodic aluminum oxide (AAO) templates with diameter in the range 30 – 70 nm. The methods and experimental conditions used to produce templates and the subsequent gold nanowires have been described in detail earlier. [19] Nanoporous gold nanowires were made in identical AAO templates of mean diameter 60 nm by electrodeposition of Au-Ag alloys, followed by dealloying the Ag using the methodology of Ji et al. [20] This results in a dense parallel array of nanoporous gold nanowires. These nanoporous nanowires are sufficiently electron transparent so that they can be imaged in the TEM without any further specimen preparation. A TEM image of an individual nanoporous gold nanowire with ligament diameter in the range 5 – 10 nm is shown in figure 1a. A diffraction pattern from the nanoporous nanowire is shown in figure 1b; this indicates that the nanoporous specimen is polycrystalline.

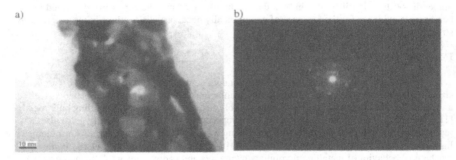

a)   b)

Figure 1 Nanoporous gold nanowires: a) Bright field TEM image showing 60 nm diameter nanoporous nanowire with ligament diameter around 5 – 10 nm. b) Diffraction pattern from one nanowire showing several grains of different orientations. (Dark spots and circled spots show two sets of diffraction patterns of (110) projection from grains with different orientations)

Uniaxial compression tests on the nanoporous gold nanowires were carried out using a Nanoindenter XP (Agilent Technologies, Santa Clara, CA, USA), fitted with a 10 µm diameter cylindrical diamond flat punch tip. The compression test procedure is identical to our earlier uniaxial compression tests on solid gold nanowire forests. [19] The load-displacement data obtained from the uniaxial compression tests are converted into engineering stress and strain for the deformation of a nanoporous nanowire (figure 2a). The onset of yield is poorly defined and is taken as the onset of deviation from linearity of the load/displacement record. The ligament

a)   b)

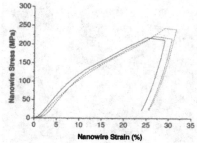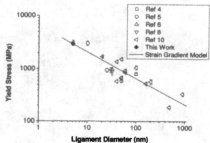

Figure 2 a) Representative stress/strain relations measured from 3 nanoporous nanowires.
b) Ligament yield stress as a function of ligament diameter for nanoporous gold from a range of
published data sources (open symbols) compared with our results (solid symbol). The solid line
shows the predicted behavior if the nanoporous gold is hardened through a strain gradient
mechanism [16].

stress is calculated from the engineering stress using the model of Gibson and Ashby. [11, 12]
From our initial alloy composition, the relative density of the nanoporous gold was estimated at
25%, which produced a ligament yield stress of about 2.5 GPa. Figure 2a also shows extensive
hardening beyond initial yield, this phenomenon was not investigated further in this study. The
computed value of the yield stress is similar to that of previous reports of the strength of
nanoporous gold with similar ligament size (figure 2b). Thus we conclude that the deformation
mechanisms that we observe in our specimens are equivalent to those in larger nanoporous gold
specimens reported in the literature.

RESULTS AND DISCUSSION

Figure 3a shows a bright field TEM image of a deformed nanoporous nanowire. A large
number of defects are observed and these are chiefly localized to, or near to, the nodes or
ligament junctions in the nanoporous foam (circled regions). This localization of deformation
structures to the nodes of the nanoporous gold is consistent with the deformation mechanism
proposed by Gibson and Ashby, [11, 12] where foam plastic collapse occurs through plastic
deformation of hinges at the nodes of the structure. A proportion of the dislocations generated in
these hinges will be retained in the structure to accommodate gradients of plastic strain
associated with the deformation; these are known as geometrically necessary dislocations
(GND). In high resolution TEM images, twins and complex microtwinned deformation
structures are seen close to the ligament nodes (Figure 3b). This indicates that the deformation of
the ligament is dominated by the nucleation of Shockley partial dislocations, and their
propagation and interaction generates the twins and complex microtwinned structures seen in the
images.

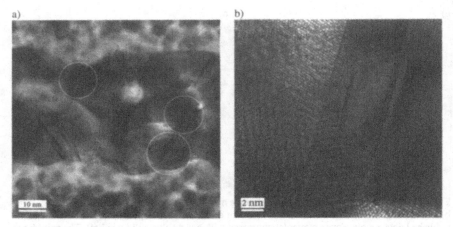

Figure 3 TEM images from deformed nanoporous nanowires prepared using method 2. a) Bright field TEM image of a deformed nanoporous gold nanowire showing defects chiefly located at the nodes between the ligaments of the structure (circled). b) High resolution TEM image showing Shockley partial dislocations and twins.

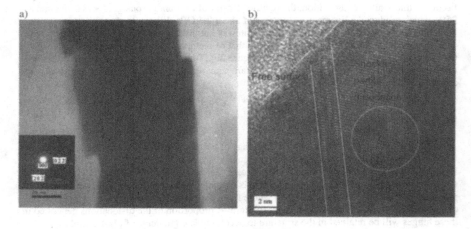

Figure 4 TEM images from solid gold nanowires deformed in compression. a) Bright field TEM image of a solid nanowire, prepared by method 1, showing characteristic surface steps and localised shear, extra spots on inset selected area diffraction pattern may indicate twinning (taken from ref. 19). b) High resolution TEM image from a specimen prepared by method 2 reveals a narrow deformation microtwin and a Shockley partial dislocation.

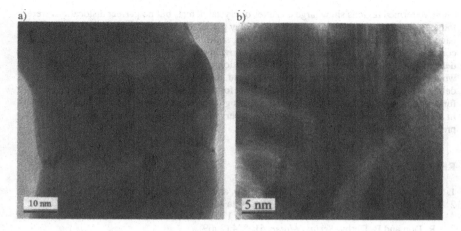

Figure 5 TEM images from a solid nanowire deformed in bending. a) Bright field TEM image showing the severe bending deformation and contrast possibly indicating deformation structures. b) Higher resolution image showing deformed region with extensive microtwins and other defects.

We have also investigated deformation structures in solid gold nanowires. In an earlier report of our work on the compression strength of metal nanowires of diameters 30 – 70 nm grown in anodic aluminum oxide (AAO) templates, TEM investigations found extensive surface steps and few, if any, dislocations. [19] This was also observed in the present study, however sparsely distributed twins and isolated Shockley partial dislocations were identified in the deformed structures (figure 4).

During the compression deformation of solid gold nanowires, the majority of the wires deform by localized shear, however occasional wires at the edge of the loaded region buckle and bend through a large angle. TEM observation of these severely deformed nanowires reveals a large defect population in the deformed region (figure 5a). A high resolution TEM image of the deformed region in a solid gold nanowire that has undergone severe bending deformation is shown in figure 5b. The deformed microstructure is similar to that found in the nodes of the deformed nanoporous gold with extensive populations of microtwins and Shockley partials.

CONCLUSIONS

We conclude that there is a significant difference between the deformation mechanisms of micropillars (or nanowires) in compression and of nanoporous gold. If plastic deformation occurs in bending (as is implicitly hypothesized by most authors working with nanoporous gold), the resulting gradients in strain must be accommodated by a population of geometrically necessary crystal defects (dislocations, twins or partial dislocations). This is observed in our nanoporous gold specimens, with deformation localized at, or close to, the nodes in the structure.

These deformed regions show large densities of crystal defects but no perfect dislocations were found, instead extensive microtwins and Shockley partial dislocations are observed. Solid gold nanowires, with diameters in the range 30-70 nm, show localized slip deformation after compression testing. The deformed nanowires show low residual dislocation density in the deformed state and this is consistent with dislocation removal at the free surfaces. However, when solid gold nanowires were deformed in bend, dislocation structures are stored in the deformed region similar to those seen with the deformation of nanoporous gold. This gives further support to our hypothesis that nanoporous gold deforms through the bending of ligaments in the structure and that part of the hardening observed in these materials arises from the presence of strain gradients.

REFERENCES

1. M. D. Uchic, D. M. Dimiduk, J. N. Florando, and W. D. Nix, *Science* **305**, 986 (2004).
2   J. R. Greer, W. C. Oliver and W. D. Nix, Acta Mater **53**, 1821 (2005); Errata: J. R. Greer, W. C. Oliver and W. D. Nix, Acta Mater **54**, 1705 (2006).
3. R. Dou and B. Derby, *Scripta Mater.* **61**, 524 (2009).
4. J. Biener, A. M. Hodge and A.V. Hamza, *Appl. Phys. Lett.* **87**, 121908 (2005).
5. J. Biener, J. et al. Nano Lett. **6**, 2379 (2006).
6. C. A. Volkert and E. T. Lilleodden, Philos. Mag. **86**, 5567 (2006).
7. C. A. Volkert, E. T. Lilleodden, D. Kramer and J. Weissmuller, *Appl. Phys. Lett.* **89**, 061920, (2006).
8. D. Lee, et al. *Scripta Mater.* **56**, 437 (2007).
9. M. Hakamada and M. Mabuchi, *Scripta Mater.* **56**, 1003 (2007).
10. A. M. Hodge, et al *Acta Mater*. **55**, 1349 (2007).
11. L. J. Gibson and M. F. Ashby, *Proc. Royal Soc. Lon. A*, **382**, 43 (1982).
12. L. J. Gibson and M. F. Ashby, *Cellular Solids: Structure and Properties*. 2nd Edn., (Cambridge University Press, 1997).
13. J. R. Greer and W. D. Nix, *Phys. Rev. B* **73**, 245410 (2006).
14. Z. W Shan, *et al. Nature Mater.***7**, 115 (2008).
15. S. H. Oh, M. Legros, D. Kiener and G. Dehm, *Nature Mater.* **8**, 95 (2009).
16. R. Dou and B. Derby, *J. Mater. Res.* In press **25**, (2010); DOI 10.1557/JMR.2010.0099.
17. Y. Sun, J. Ye, A. W. M. Minor and T. J. Balk, *Microscopy Res. Tech.* **72**, 232, (2009).
18 Jin, H. J. *et al. Acta Mater.* **57**, 2665 (2009).
19. R. Dou and B. Derby, *Scripta Mater.* **59**, 151 (2008).
20. C. X. Ji and P.C. Searson, *J. Phys. Chem.* **B, 107**, 4494 (2003).

Mater. Res. Soc. Symp. Proc. Vol. 1224 © 2010 Materials Research Society   1224-FF10-10

Elastic Properties of Mimetically Synthesized Model Nanoporous Carbon

Xi Mi[1] and Yunfeng Shi[2]
[1] Department of Physics, Applied Physics and Astronomy, Rensselaer Polytechnic Institute
[2] Department of Materials Science and Engineering, Rensselaer Polytechnic Institute,
Troy, New York 12180, USA.

ABSTRACT
Activated carbon is widely used for its attractive diffusion, adsorption and reaction properties. However, its mechanical behavior has received much less attention. We present a molecular dynamics simulation study on the elastic properties of activated carbon with nanometer-sized pores. The nanoporous carbon sample is composed of curved and defected graphene sheets, which is synthesized using quench molecular dynamics (QMD) method [1]. One unique feature of the current model is the mechanical stability, thus the bulk modulus, Young's modulus, shear modulus and Poisson's ratio can be obtained from simulated mechanical tests. By varying the density of the nanoporous carbon model, it was further found that the bulk modulus vs. density relation follows Gibson-Ashby type power law with exponents of 2.80 at low densities and 1.65 at high densities.

INTRODUCTION
Nanoporous carbon materials have drawn substantial research interests because of their unique properties: extremely large surface area, preferential adsorption and the potential to be used as catalysts for chemical reactions [2]. A prerequisite for applications that fully exploit the above properties of these materials is the knowledge of their mechanical properties. Although indentation tests have been conducted on various nanoporous carbon materials [3-6], the systematic experimental or theoretical studies on mechanical properties remain absent. On the other hand, mechanical properties of cellular materials have been established in materials with macroscopic pores [16], while the typical size in nanoporous carbon is between a few and a few tenths of nanometers. Whether the structure-property relation in foams is valid for systems with nanometer-sized pores has not been fully examined. Nanoporous carbon can be seen as a model system to provide insights on the structure-property relation for nanoporous materials in general.

Empowered by the rapid growth of computation power, molecular simulation techniques have provided valuable atomic information on the porous structure of nanoporous carbon, adsorption and diffusion behavior of its guest species [2, 7-10]. However, there is very few viable atomic level nanoporous carbon model that are both structurally realistic (i.e. agrees with experimental structural signatures) and mechanically stable (thus can be subjected to mechanical tests). The widely used slit pore models [11] fail to match the structural signatures. Recent hybrid reverse Monte Carlo (HRMC) method [12, 8] is able to generate atomic models with structural signatures that match the experiments. However, the structural signature deviates rapidly from the correct values upon thermal relaxation which indicates that the atomic models from HRMC are mechanically unstable and can only be used with the atomic positions fixed.

Recently, nanoporous carbon structures are obtained using quench molecular dynamics (QMD) simulation [1]. Excellent agreement has been achieved with saccharose-based activated carbon CS1000a [7] in terms of density, chemical composition and pair-correlation function. Importantly, this virtual synthesis process enables the production of mechanically stable structures. Nonetheless, there are two main shortcomings with the prior QMD simulations. First, the final temperature is 6,600 K, far above the temperature under which most indentation tests are conducted in experiments. Second, five member carbon rings are excluded by the strong angular constraints imposed by the force field (although seven member and larger rings do exist). As carbon pentagons exist in other carbon nanomaterials such as fullerenes as well as nanoporous carbon models from HRMC, the exclusion of carbon five member rings may be unphysical.

In this work, an improved sample generation procedure is provided. The first modification is that the final temperature after quenching is changed to 290 K instead of the original 6,600 K. The second modification is to impose a varying angular constraint, which allows carbon pentagons to grow spontaneously into the porous network. In addition, we employ the structure factor $S(q)$ as the primary characterization means to scrutinize simulation models against the experiments, which is shown to be superior than the pair correlation function $g(r)$. Using the improved nanoporous carbon model, the elastic properties including bulk modulus, Young's modulus, shear modulus and Poisson's ratio are calculated from molecular dynamics (MD) simulation. By producing nanoporous carbon at different densities, we further studied the scaling behavior of the bulk modulus as a function of density.

SAMPLE PREPARATION

We use reaction state summation (RSS) method for carbon potential calculation [1, 13]. The chief merit of this force field is its capability of describing covalent bond breaking and forming with moderate computational demands. It contains a two-body potential term to determine the bond length, a three-body potential term to restrict the bond angles and a short-range two-body potential term to prevent atoms from overlapping. The final form is the summation of those three terms each modulated by a coordinate-dependent weight function. Only sp^2 hybridization state of carbon is considered here due to the fact that CS1000a has dominating graphene-like structural signatures. The domination of sp^2 is also observed in HRMC simulations [8]. The three body potential is determined by an angular constraint function which can be written (without its cutoff functions) as,

$$E^A(\theta) = \varepsilon G(\cos\theta - \cos 120°)$$ (1)

where θ is the bond angle, ε is an energy parameter and $G(x)$ is a piecewise function composed of three quadratic segments and two linear segments (details in Ref. [1]).

QMD simulation is carried out for the sample generation following similar procedure described in Ref. [1] except for the improvements discussed below. The system starts initially from free carbon gas at a high temperature ($T_{initial}$=15,000 K) and then it is quenched gradually down to final temperature (T_{final}=290 K) in the NVT ensemble with an initial density of 0.038 atoms/Å^3 or 0.757 g/cm^3. This procedure is designed to mimic the pyrolysis process after the polymer chains break down and species other than carbon are evaporated. The timestep for integrating the equations of motions for MD is 0.07 fs. Snapshots of the porous network structure during the growth process are provided in Fig. 1. The sample remains a box of mostly monatomic carbon atoms until 10,000 K. The majority of the porous carbon network is formed from 10,000 K to 7,700 K.

The comparison of the sample preparation procedure between this work and the previous work [1] is shown in Table 1. First of all, the angular constraint function is modified so that bond angles have a broader distribution to accommodate the 108° angle of carbon pentagons. The angular range of stable bond angle θ is given by the linear region in $G(x)$ which is determined by

$$\arccos(a_2 + \cos 120°) < \theta < \arccos(a_2 - \cos 120°) \tag{2}$$

where a_2 is a constraint parameter. In the previous work, a_2 is chosen as 0.08 which restricts θ between 115° and 125°. Here a_2 is linearly changing with time from 0.08 to 0.2 during the quenching so that bond angles can range from 107° to 143° in the product. As can be seen in Table 2, the current scheme produces more 7-, 8- and 9- member rings than those produced in Ref. [1]. More importantly, carbon pentagons that are prohibited before can now form spontaneously. Nonetheless, as in both procedures, carbon hexagons dominate with identical densities. Therefore, the general porous network should be structurally similar.

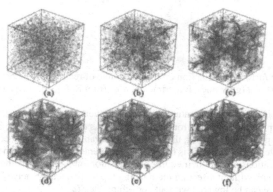

Figure 1. Snapshots for atom configuration of the sample during the quenching. (a) 10,300 K (b) 10,000 K (c) 9,600 K (d) 9,100 K (e) 7,700 K (f) 290 K

Table 1. List of the differences of sample generation procedures in previous [1] and this work

Sample	T_{final} (K)	Quench time (ps)	Box size (Å)	$N_{initial}$	N_{final}	a_2
Previous work	6,600	3200	37.77	2048	1957	0.08
This work	290	5600	94.43	32000	32000	0.08 to 0.2

Table 2. Ring distributions per atom for the final structure simulated. Ring distributions from the previous work [1] are listed for comparison. There are no smaller rings than pentagons.

Ring Size	Five	Six	Seven	Eight	Nine
Previous work	0	0.375	0.0265	0.00460	0.00307
This work	0.0000938	0.343	0.0456	0.0178	0.00865

Secondly, we choose to scrutinize the structure of the simulated nanoporous carbon structure in terms of the structure factor $S(q)$ instead of the pair correlation function $g(r)$. $S(q)$ in our simulation system is directly calculated from atomic positions and yields excellent agreement with that of the experiments of the saccharose based activated carbon CS1000a [7] of the same density, as is shown in Fig. 2. The simulated curve reproduces the positions, heights and widths of all six major peaks reasonably well. The justification of using $S(q)$ instead of $g(r)$ is as follows. Unlike simulation models in which both $S(q)$ and $g(r)$ can be readily calculated, the experimental $g(r)$ is difficult to measure directly and is often obtained by transforming the experimental $S(q)$ to real space. However, such conversion requires high quality scattering data at large wave vectors which may not be available. Instead, by comparing to the experimental $S(q)$ directly, the conversion process from reciprocal space to real space is obviated.

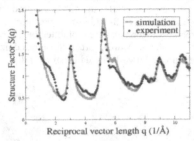

Figure 2. Structure factors as a function of absolute value of reciprocal vectors for the atomic model here for CS1000a (gray line). Experimental $S(q)$ from Ref. [7] is also shown (black dots).

Thirdly, 290 K is the final temperature at the end of quenching. In contrast, the final temperature of the previous synthesis routine is 6,600K. As can be seen in Fig 3, $g(r)$ is very sensitive to the temperature. However, there are no observable differences in corresponding $S(q)$ curves at different temperatures. This is because the temperature determines the amplitude of atomic thermal vibrations around their equilibrium positions and modulates the shapes of the peaks in the $g(r)$ graph without affecting the bonding topology once the network is formed. Thus, the final temperature can be lowered without perturbing the $S(q)$.

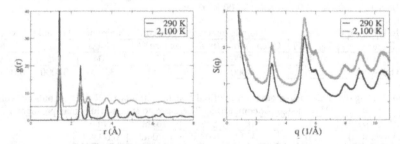

Figure 3. Pair correlation functions (left) and structure factors (right) for same sample under different temperatures are plotted. Both graphs are shifted for clarity.

Lastly, a much larger system with a cubic box with 94.43 Å in each side and a total of 32000 atoms are simulated as compare to 37.77 Å and 1957 atoms in the previous work and 25 Å and 566 atoms in HRMC [8] for the same target sample CS1000a. Note that the sample generation procedure in this work leaves no carbon atoms in the gas phase at the final temperature. The number of atoms in the simulation system remains the same before and after the quenching ($N_{initial} = N_{final}$). The ability to prepare molecular nanoporous models with different sizes permits later studies on the system size-effect of adsorption, diffusion and mechanical properties of nanoporous carbon and samples with large pores.

ELASTIC PROPERTIES

The as-prepared sample undergoes a hydrostatic relaxation under NPT ensemble to eliminate the residue stress. The simulation box resizes to a cubic with 93.22Å on each side. The density then changes about 4% to 0.0395 atoms/Å^3 or 0.787 g/cm^3. Hydrostatic and uniaxial tensions as well as shear deformations are then performed on the sample by MD simulations using RSS Carbon force field under NVT ensemble with temperatures equal 0.01 K to suppress thermal fluctuations. The strain rates are set to constants equal to 4.29×10^9 /s in hydrostatic tests, 1.43×10^9 /s in uniaxial tests and 1.51×10^9 /s in shear tests. These strain rates are low enough based on the observation that lower strain rates do not lead to observable differences in stress-strain curve, potential energy or any of the elastic moduli. Young's modulus of 32 GPa, bulk modulus of 17 GPa and shear modulus of 14.3 GPa are obtained by linear fitting in the low loading region of the stress-strain curve. Poisson's ratio can be calculated from bulk modulus and Young's modulus to be 0.19.

Although RSS Carbon force field is successful in capturing the correct structural signatures, its usage in simulating mechanical tests is not well examined before. AIREBO [14] is a well tested empirical potential field for carbon and hydrocarbon systems. It is used here as a benchmark for our mechanical tests. We choose a much smaller sample to reduce the computing time. The bulk modulus of the small sample is 14.5 GPa which is close to the large sample. The sample with 2048 atoms is relaxed then subjects to a hydrostatic tension using AIREBO. Bulk modulus is 13.0 GPa under this potential field. The excellent agreement between these two force fields verifies the use of RSS Carbon potential in the mechanical tests in the elastic regime.

To see how density affects bulk modulus, 14 samples with different densities are prepared. We start with 2048 atoms in each sample. All samples are prepared using the same procedure described above. The bulk moduli of all these samples are calculated. As is shown in Fig.4, the data at low density (< 0.08 atoms/Å^3) can be well fit into a scaling law: $K \sim \rho^{2.80}$. This is in agreement with the scaling power law with an exponent of 2.69 measured in carbon aerogel and xerogels [15]. The high density region, however, can fit into $K \sim \rho^{1.65}$.

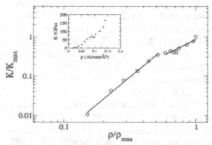

Figure 4. A log-log plot of bulk modulus scaling with density normalized by the bulk modulus and density of the densest sample here. Circles denote simulation data. Straight lines are the exponential fittings at different density regimes. Inset: the corresponding linear-linear plot.

This piecewise scaling power law suggests there are two different mechanisms that are responsible for the elastic properties in different density regimes. At low density, the scaling exponent between 2 and 3 suggests that its elastic deformation mechanism is cell-wall bending, similar to that of cellular materials such as foams [16] although these two materials have significant differences in pore sizes. It should be noted that the Gibson-Ashby power law usually has a constant exponent over a wide range of densities. The transition of scaling power law here between low densities and high densities cannot be solely explained by Gibson-Ashby's cellular model. One plausible explanation is that at high densities, parts of the carbon atoms branch out of the load-bearing network thus do not contribute to the bulk modulus nor carry the force chains [17]. Therefore, the density increase due to the formation of these branches leads to little bulk modulus enhancement. It is conjectured that a proper effective density is superior to the apparent density in understanding the scaling behavior. Research along this line is underway.

CONCLUSIONS

Nanoporous carbon sample is obtained by QMD using an improved RSS Carbon force field. This sample matches the result from scattering experiments and is mechanically stable. We obtained bulk modulus, Young's modulus and shear modulus and Poisson's ratio using MD simulations. Gibson-Ashby scaling relationship is also observed for nanoporous carbon at different densities: bulk modulus increases with the density in a power law form with an exponent of 2.8 at low densities and of 1.65 at high densities.

ACKNOWLEDGEMENT

We thank Mr. Jeremy Palmer and Professor Keith Gubbins at North Carolina State University for useful discussions. MD simulations are performed using LAMMPS with customized force fields. Simulations were carried out in the Computational Center for Nanotechnology Innovations (CCNI) at Rensselaer Polytechnic Institute.

REFERENCES

1. Y. F. Shi, *J. Chem. Phys.* **128**, 234707 (2008).
2. M. J. Biggs and A. Buts, *Mol. Simul.* **32**, 579 (2006).
3. S. Field and M. V. Swain, *Carbon* **34**, 1357 (1996).
4. M. Sakai, H. Hanyu, and M. Inagaki, *J. Am. Ceram. Soc.* **78**, 1006 (1995).
5. N. Iwashita and M. v. Swain, *Mol. Cryst. Liq. Cryst.* **386**, 39 (2002).
6. D. A. Ersoy, M. J. McNallan, and Y. Gogotsi, *Mat Res Innovat* **5**, 55–62 (2001).
7. S. K. Jain, J. P. Pikunic, R. J.M. Pellenq, and K. E. Gubbins, *Adsorption* **11**, 355 (2005).
8. S. K. Jain, R. J.M. Pellenq, J. P. Pikunic, and K. E. Gubbins, *Langmuir* **22**, 9942 (2006).
9. L. D. Gelb and K. E. Gubbins, *Langmuir* **14**, 2097 (1998).
10. L. D. Gelb and K. E. Gubbins, *Langmuir* **15**, 305 (1999).
11. P. H. Emmett, *Chem. Rev.* **43** 69 (1948).
12. T. Peterson, I. Yarovsky, I. Snook, D. McCulloch and G. Opletal, *Carbon* **42**, 2457 (2004).
13. Y. F. Shi and D. W. Brenner, *J. Chem. Phys.* **127**, 134503 (2007).
14. S. J. Stuart, A. B. Tutein and J. A. Harrison, *J. Chem. Phys.* **112**, 6472 (2000).
15. J. Grob and J. Fricke, *NanoSturctured Materials.* **6**, 905–908 (1995).
16. L. J. Gibson and M. F. Ashby, *Cellular solids*, 2nd ed. (Combridge University Press, 1997) pp. 175-231.
17. J. F. Peters, M. Muthuswamy, J. Wibowo and A. Tordesillas, *Phys. Rev. E* **72**, 041307(2005).

Mater. Res. Soc. Symp. Proc. Vol. 1224 © 2010 Materials Research Society   1224-FF10-35

Microstructural and Mechanical Properties of Boron Carbide Ceramics by Methanol Washed Powder

Jong Pil Ahn[1,2] , Kyoung Hun Kim[1], Joo Seok Park[1], Jae Hong Chae[1], Sung Min So[1,2] and Hyung Sun Kim[2]
[1] Korea Institute of Ceramic Engineering and Technology, 233-5, Gasan-Dong, Guemcheon-Gu, Seoul, 153-801, Korea
[2] Information Materials Lab. Materials Engineering, Inha University, 253, Yonghyun-dong, Nam-gu, Incheon, 402-751, Korea

ABSTRACT

B_4C ceramics were fabricated by a hot-press sintering and its sintering behavior, microstructure and mechanical properties were evaluated. The relative density of B_4C ceramics were obtained by a hot-press sintering reached as high as 99% without any sintering additives. The mechanical properties of B_4C ceramics was improved by a methanol washing which can remove B_2O_3 phase from a B_4C powder surface. This improvement is resulted from the formation of homogeneous microstructure because the grain coarsening was suppressed by the elimination of B_2O_3 phase. Particularly, the mechanical properties of the sintered samples using a methanol washed powder improved compared with the samples using an as-received commercial powder.

INTRODUCTION

Boron carbide (B_4C) is very useful ceramics in engineering applications because of its excellent mechanical properties. B_4C is premier material for personal ballistic armor because it is the third hardest material behind diamond and cubic boron nitride (c-BN), in addition to its low theoretical density (2.52 g/cm3) [1]. B_4C is also used as abrasive applications such as abrasive powder and nozzles due to its excellent abrasion and wear resistance [2]. It has been reported that the mechanical properties of B_4C ceramics depend on the B/C ratio because Boron Carbide exists as a solid solution in the range of 8.8 to 20 at% carbon ($B10.5C - B_4C$) [3 - 5]. It was reported that the hardness and fracture toughness values were maximum at essentially the stoichiometric composition B/C = 4 and decreased for B/C < 4 (composition region of B_4C + C) and for B/C > 4 (non-stoichiometric region of B_4C) [5].

The B_4C has low specific gravity and mechanical properties to use lightweight armor and the high temperature material. Densification of B_4C is difficult, because of high covalent bonding and boron oxide (B_2O_3) film on B_4C particles surface which can cause a microstructural coarsening during sintering. Sintering behavior, microstructure and mechanical properties of hot press sintered B_4C were evaluated.

EXPERIMENTAL DETAILS

Commercially available B_4C powder (HS grade, H.C. Starck, Germany) was used as a starting material. The commercial B_4C powder was suspended in methanol (99.9%, J.T. Baker, USA) and ultrasonicated (B3200, Branson, USA) for 1 hour. The suspension was then placed in dry oven at 80°C for 12 hours to volatilize methanol. This process was repeated three times for the purpose of removing B_2O_3 coatings from the surface [6].

The powders were loaded into a graphite mold and then Hot press sintering(MHP-350, Mono Cerapia, Korea) at temperature range from 1900°C to 2050°C for 1 hour at pressure of 40MPa in a vacuum. Density was measured by the Archimedes method and X-ray diffractometer

(M03XHF22, MAC Science Co., Japan) was used for analyzing the phases in powders and sintered bodies. Hardness was measured using a Vickers indenter (HM-124, Akashi, Japan) with a 9.8 N load for 15 seconds. The microstructure characterization was carried out with Field Emission Scanning Electron Microscope (FE-SEM, 6700F, JEOL, Japan) after electrolytic etching in 1% KOH solution for 30 seconds with 3.5 V and 0.1 A/cm2. Average grain size was calculated using the linear intercept method (ASTM E 112-96) based on 1000 measurements on SEM micrographs.

RESULT AND DISCUSSION

The XRD patterns of B_4C powder and sintered body at 2000°C indicates that the methanol washing process is very effective in removing B_2O_3 coatings from the B_4C powder surface. Also, B_2O_3 peak was removed from the sintered body using a commercial powder because B_2O_3 phase has a tendency to vaporization over 1200°C.

The relative density of methanol washed specimens dramatically increased at 1950°C up to 98%, and it reached over 99% at 2000°C but non-washed specimens constantly increased at 2050°C. This result indicates that the specimen of methanol washing process has better effective sinterbility than non-washed specimens.

The average grain size of the commercial powder specimens was larger than that of methanol washed powder specimens. In general, the grain coarsening of covalently bonded materials originates from surface diffusion and evaporation-condensation mechanism. It was suggested that the existence of B_2O_3 coatings may provide a rapid diffusion path on particle surfaces because the firing temperatures are well above its melting point, 450°C. Also, the volatility of B_2O_3 causes the microstructural coarsening above 1500°C because the vapor pressure of B_2O_3 is sufficiently high for mass transport [7]. Thus the retardation of densification and the microstructural coarsening are due to the existence of B_2O_3 phases on the particles surface.

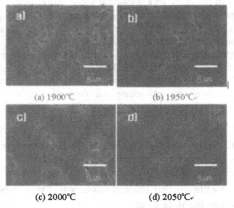

(a) 1900°C   (b) 1950°C

(c) 2000°C   (d) 2050°C

Fig. 1 SEM micrographs of etched surfaces B_4C specimens sintered at (a) 1900°C, (b) 1950°C, (c) 2000°C and (d) 2050°C using commercial powder.

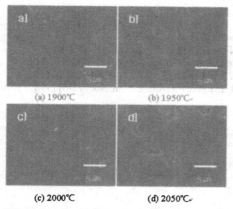

(a) 1900°C   (b) 1950°C

(c) 2000°C   (d) 2050°C

Fig. 2 SEM micrographs of etched surfaces B_4C specimens sintered at at (a) 1900°C, (b) 1950°C, (c) 2000°C and (d) 2050°C using methanol washed powder.

Figure 1 and figure 2 show the etched surface morphology of sintered bodies with sintering temperature using the commercial B_4C powder and the methanol washed B_4C powder, respectively. Near full dense microstructures are shown in both specimens over 2000°C. The grain coarsening during Hot press sintering is clearly seen the commercial powder specimen (as shown in Fig. 1) with increase sintering temperature. However, the microstructure of methanol washed specimen (as shown in Fig. 2) is finer and more homogeneous than that of commercial powder specimen. Referring to the result of average grain size, these results show that the removal of B_2O_3 coatings is very effective to restrict the grain coarsening and consequently to obtain a fine and homogeneous microstructure.

The increase of Vickers hardness values of both samples with increase sintering temperature up to 2000°C may originated from the increase of density. The Vickers hardness value of methanol washed specimens is higher than commercial specimens because the methanol washed specimens have a smaller grain size. Fine and homogeneous microstructure make a great contribution to improvement of the mechanical properties.

CONCLUSION

Poorly sinterable B_4C ceramic were successfully fully densified even though without additives. The microstructural coarsening was effectively inhibited by the removal of Boron Oxide coatings on B_4C particles surface by methanol washing process. The removal of B_2O_3 coatings could permit a direct B_4C - B_4C contact resulted in low temperature densification and suppression of coarsening by the acceleration of lattice and / or grain boundary diffusion. As a result, the mechanical properties were significantly improved by the formation of a fine and homogeneous microstructure.

REFERENCES

1. F. Thevenot, J. Euro. Ceram. Soc. 6[4] (1990) 205-225.
2. D. Simeone, C. Mallet, P. Dubuisson, G. Baldinozzi, C. Gervais, and J. maquet, J. Nucl.

Mater. 277 (2000) 1-10.
3. J. Beauvy, J. Less Common Metals 90[2] (1983) 169-175.
4. R. D. Allen, J. Am. Chem. Soc. 75[14] (1953) 3582-3583.
5. K. Niihara, A. Nakahira, and T. Hirai, J. Am. Ceram. Soc. 67[1] (1984) c13-c14.
6. P. D. Williams, and D. D. Hawn, J. Am. Ceram. Soc. 74[7] (1991) 1614-1618.
7. S. L. Dole, S. Prochazka, and R. H. Doremus, J. Am. Ceram. Soc. 72[6] (1989) 958-966.

Mater. Res. Soc. Symp. Proc. Vol. 1224 © 2010 Materials Research Society   1224-FF10-17

Temperature Variation in Energy Absorption System Functionalized by Nanomaterials

Yu Qiao,[1] Zhongyuan Sun,[2] Weiyi Lu,[1] and Aijie Han[2], *
[1]Department of Structural Engineering, University of California-San Diego, La Jolla, California 92093, USA
[2]Department of Chemistry, University of Texas-Pan American, Edinburg, Texas 78539, USA

ABSTRACT

The thermal effect on the nanofluidic behaviors in a nanoporous silica gel is investigated experimentally. When a nanoporous silica gel is modified by silyl groups, its surface becomes hydrophobic. A sufficiently high external pressure must be applied to overcome the capillary effect; otherwise liquid infiltration could not occur. The formation and the disappearance of a solid–liquid interface are employed for energy storage or dissipation. When the hydrophobic surface of nanoporous silica gel is decomposed at various temperatures, the organic surface layers can be deactivated. As a result, the degree of hydrophobicity, which can be measured by the liquid infiltration pressure, is lowered. The infiltration and defiltration behaviors of liquid are dependent on the controlled by the decomposition-treatment temperature.

INTRODUCTION

Porous materials have been widely applied in chemistry, biology, and energy-related areas [1]. Depending on their synthesis and treatment process, the nanopores inside of the structures can be either close-end or open-end, and the pore shape can be either ordered or disordered. Besides the pore diameter and shape, the component of the nanoporous materials can also affect their property and application. Recently, a system consisting of lyophobic nanoporous materials in liquid phases to produce the colloidal suspension has attracted the interest of experts in terms of their applications as mechanical actuators, thermal machines, or dampers [2-3].

The smart systems are developed by dispersing lyophobic nanoporous materials in liquid phases. Applying an external pressure, the liquids overcome the capillary effect and enter the nanopores. During this process, a significant amount of external work is converted to the excess solid/liquid interfacial tension, $\Delta\gamma$. One way to calculate the specific absorbed energy is E= $\Delta\gamma$ A, where A is the specific surface area of the nanoporous material. Typically, $\Delta\gamma$ is 10-50 mJ/m^2 [4]. Since the surface area of nanoporous materials is as large as ranging from 100 to 1000 m^2/g, E can be 1-50 J/g, attractive for developing lightweight and small-sized protective and damping devices, e.g. soldier armors and vehicle bumpers [5].

During the past a few decades, a number of nanoporous materials have been developed [6-7], among which nanoporous silica gel is another good candidate for the energy absorption system. Their nanopore size can be controlled in the range of 2-100 nm and their specific surface areas are around 100-1000 m^2/g [8]. Furthermore, the nanoporous structure of silica gel is quite stable even under a high pressure of around 500 MPa. Usually, nanoporous silica gel is obtained through the aggregation of nanograins. Accordingly, the outer as well as the inner surfaces of the silica gel are covered by rich silanol groups (OH groups). Therefore, the raw nanoporous silica gel is hydrophilic and naturally penetrated by water and needs to be grafted with hydrophobic layers before using in the new systems. Since the silanol groups are the active sites to react with the modification agents, the more silanol groups before the modification, the better coverage will be after the modification. As shown in previous reports [9-10], the degree of hydrophobicity of silica gel which is related to the surface coverage of silyl groups, c, dominates the infiltration pressure and the absorbed energy density. The value of c can be controlled quite precisely by surface treating nanoporous silica for different periods of time. However, such a technique may not work well if the required surface group density is relatively low. If the treatment time is too short, the diffusion of silyl molecules inside nanopores cannot reach the equilibrium condition, and the surface coverage may be nonuniform along the axial direction. To overcome the disadvantage, we investigate another method to control the surface coverage: decomposition-treatment. After a saturated surface group layer is produced, its density can be lowered by heating the material. Since the decomposition process is directly related to the temperature field and the effects of molecular diffusion should be secondary, the change in surface properties tends to be homogeneous.

EXPERIMENT

The porous silica gel was obtained from Aldrich (#60758), with the modal value of nanopore size of 7.0 nm and the specific nanopore surface area of 234 m^2/g. The material characterization was performed by using a TriStar-3000 gas absorption analyzer. The as-received material was in powder form modified with chloro(dimethyl)octadecylsilane, with the particle size in the range of 20-50 μm. Following vacuum drying at 100 °C for 24 h, the silica gel was mixed with 40 mL of dry toluene and 0.5 mL of chlorotrimethylsilane at 90 °C for 5 days. The mixing rate was about 20 rpm. During this process, the small chlorotrimethylsilane molecules diffused into the nanopores and deactivated the hydroxyl groups left even after chloro(dimethyl)octadecylsilane modification, forming $OSi(CH_3)_3$ groups [11]. Due to the relatively long treatment time, the surface group density was saturated; that is, further increase in treatment time would not lead to detectable variation in the degree of hydrophobicity [12]. The end-capping treated silica gel was collected in a vacuum filter, washed by dry toluene and warm water for a few times, and dried in vacuum at 100 °C for 5 h. The average nanopore diameter of the surface treated silica gel has been found to be 6.0 nm, and the specific surface area was 219 m^2/g. The material was then placed in a tube furnace and heated at 180 °C. The treatment time was ranged from 0 to 75 min. The setpoint of 180 °C was chosen because a higher or lower temperature would lead to a reduced controllability of treatment time. An air flow was maintained during the decomposition-treatment.

The surface property of the decomposition-treated silica gel samples was examined through a pressure-induced liquid infiltration experiment. The mixture of silica gel and water was sealed in a stainless steel container by a piston. By using a material testing machine, Instron 5569, the piston was compressed into the container at a constant rate of 1 mm/min, resulting in the external load F and the piston displacement δ. The liquid pressure and the liquid volume change were calculated as $P=F/A_0$ and $\Delta V= \delta A_0$, respectively, where A_0=287 mm^2 is the cross-sectional area of the container. When the infiltration was completed, the piston was moved out at the same rate.

DISCUSSION

Typical sorption isotherm curves for the nanoporous silica gel after various decomposition-treatment times are analyzed through a pressure-induced liquid infiltration experiment shown in

Fig.1. The dashed line in Fig.1 shows the behavior of the silyl modified silica gel without any decomposition-treatment. Another three solid curves are for the surface modified silica gel after decomposition-treatment for 30, 50 and 75 min, respectively.

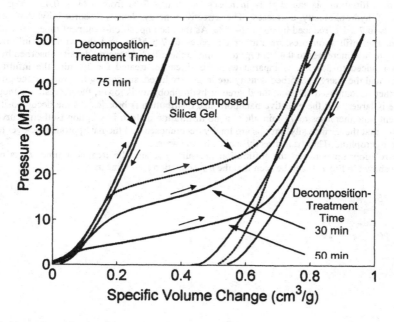

Fig.1. Typical sorption isotherm curves at room temperature.

As the nanopore surface is modified by silyl groups, soaking does not spontaneously occur when the hydrophobic nanoporous silica gel is immersed in water at ambient temperature. Based on the well-known capillary phenomenon, external work must be done to spread a non-wetting liquid on a solid surface. As an external pressure is applied, the liquid is forced to infiltrate into the energetically unfavorable hydrophobic nanopores. The critical pressure point is called infiltration pressure, p_{in}, taken as the pressure at the middle point of the infiltration plateau. As the liquid starts to enter the nanopores, the system volume largely decreases with a relatively small pressure increasing. During this process, a significant amount of external work is converted to the excess solid-liquid interfacial tension. For the surface modified silica gel without any decomposition-treatment, the rapid change in volume causes the formation of the infiltration plateau in sorption isotherm curve around 22 MPa. The specific absorbed energy can be calculated as $E = P \, \Delta V$, where P is the infiltration pressure and ΔV is the system volume change that results from the nanopore volume change. Note that the energy density absorbed by the system is as high as 9.5 J/g for the silyl modified silica gel without any decomposition-treatment.

After the decomposition-treatment for 30 min at 180 °C, the property of silica gel changes. A certain portion of the organic surface groups decomposes, so that less nanopore wall area is covered by the hydrophobic layer. Though the sorption isotherm curve remains similar with the one without decomposition-treatment, the infiltration pressure decreases to 16 MPa, and the width of infiltration plateau slightly increases by about 4.7 % from 0.43 to 0.45 cm³/g. The degree of hydrophobicity decreases. The effective energy density absorbed by the system decreases to 7.2 J/g, reduced by about 24.2 %. As the decomposition-treatment time increases to 50 min, the infiltration pressure further decreases to 7.6 MPa, and the width of infiltration plateau slightly increases to 0.47 cm³/g, resulting in the effective energy density absorbed by the system reduced to 3.6 J/g. Compared with the silica gel treated for 30 min, the infiltration pressure and the energy absorbed density are largely reduced, suggesting that the surface groups are further decomposed. That is, the degree of hydrophobicity is lower; the accessible nanopore volume is larger; and the effective nanopore size distribution is broader. As the decomposition-treatment time increases to 75 min, there is no infiltration plateau in sorption isotherm curve. It indicates that the surface silyl groups are totally decomposed and the nanoporous silica gel turns back to hydrophilic. Thus, it is directly soaked by the water.

The relationship between the infiltration pressure, p_{in}, and the treatment time, t_d, is more clearly shown in Fig.2. It can be seen that the p_{in}–t_d relation is nonlinear.

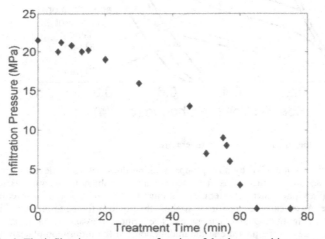

Fig. 2. The infiltration pressure as a function of the decomposition-treatment time.

When t_d is relatively small, its influence is mild. When the treatment time increases from 0 to 30 min, the infiltration pressure is reduced from 22 MPa to 16 MPa by 27.3 %. However, when the treatment time increases from 30 min to 60 min, the infiltration pressure tends to be 0. Since the particle size is quite small, the temperature field should reach the equilibrium condition rapidly. For t_d longer than a few min, the p_{in}–t_d relation should be dominated by the thermally aided decomposition process. As we know, the silyl groups are attached to the nanopore wall through the oxygen atom, if the O-Si bond is broken, further breakage of Si-C bonds from the

76

silyl groups themselves would not affect the surface property of the silica gel. Moreover, the influences of the end group and the side group can be different. When the decomposition treatment time is longer than 1 h, most of the silyl groups are deactivated and the nanopore wall becomes hydrophilic. Under this condition, no infiltration of liquid can be observed, since the nanopores are filled spontaneously immediately after the silica gel is immersed in the liquid phase.

CONCLUSIONS

Effects of decomposition-treatment temperature on energy absorbed density of a surface modified nanoporous silica gel is inverstegated in the work. The influence from the surface group coverage and the decomposition-treatment temperature are analyzed by pressure-induced infiltration experiments. As the silyl modified nanoporous silica gel is heated, the surface group coverage and the degree of hydrophobicity are lowered with the decomposition temperature increasing, resulting in a decreasing of energy absorption density.

ACKNOWLEDGMENTS

This work was supported by The Army Research Office under Grant No. W911NF-05-1-0288.

REFERENCES

[1] S. Polarz and B. Smarsly: Nanoporous Materials. *J. Nanosci. Nanotechnol.* 2, 581 (2002).
[2] Martin T., Lefevre B., Brunel D., et al., Dissipative water intrusion in hydrophobic MCM-41 type materials. *Chem. Commun.*, 24 (2002).
[3] Han A., Punyamurtula V.K. and Qiao Y. Effects of cation size on infiltration and defiltration pressures of a MCM-41. *Appl. Phys. Lett.*, 92, 153117, 1 (2008).
[4] J. C. Riviere and S. Myhra: *Handbook of Surface and Interface Analysis.* CRC Press (1998).
[5] G. Lu and T. Yu: *Energy Absorption of Structures and Materials.* CRC Press (2003).
[6] Karami and S. Rohani: Progressive strategies for nanozeolite Y synthesis: A review. *Rev. Chem. Eng.* 23, 1 (2007).
[7] Zukal: Recent trends in the synthesis of nanoporous materials. *Chem. Listy* 101, 208 (2007).
[8] M. Yang and K. J. Chao: Functionalization of molecularly templated mesoporous silica. *J. Chinese Chem. Soc.* 49, 883 (2002).
[9] Han and Y. Qiao: Controlling infiltration pressure of a nanoporous silica gel via surface treatment. *Chem. Lett.* 36, 882 (2007).
[10] Han and Y. Qiao: Effects of surface treatment of a MCM-41 on motions of confined liquids. *J. Phys. D – Appl. Phys.* 40, 5743 (2007).
[11] M. H. Lim and A. Stein: Comparative studies of grafting and direct syntheses of inorganic-organic hybrid mesoporous materials. *Chem. Mater.* 11, 3285 (1999).
[12] J. A. Fay: *Fluid Mechanics.* MIT Press (1994).

Polymers & Composites

Mater. Res. Soc. Symp. Proc. Vol. 1224 © 2010 Materials Research Society   1224-FF11-10

Anisotropic Strain and Training of Conducting Polymer Artificial Muscles under High Tensile Stresses

Keiichi Kaneto, Hikaru Hashimoto, Kazuo Tominaga, Tomokazu Sendai and Wataru Takashima

Graduate School of Life Science and Systems Engineering, Kyushu Institute of Technology, 2-4 Hibikino, Wakamatsu-ku, Kitakyushu 808-0196, Japan

ABSTRACT

Electrochemomechanical deformations (ECMD) of conducting polymer, polyaniline, films are studied to investigate the creeping and the memory effects. During electrochemical cycling under high tensile stresses up to 5 MPa, the films showed a remarkable creeping, resulting in the one dimensional anisotropic deformation. However, the creeping was recovered by release of the tensile stress, restoring from the anisotropic deformation. It was also found that the strain of ECMD after applying high tensile stresses increased compared with that before applying the large tensile stress. The result indicates that the artificial muscles are strengthened in strain by the experience of large tensile loads, and discussed taking the rheology of electrochemical cycles, viz., electrostatic crosslinking of polymer chains by oxidation and release of crosslinking by reduction.

INTRODUCTION

Conducting polymers deform by electrochemical oxidation and reduction, resulting from the insertion of bulky ions as well as the change of polymer conformation due to the delocalization of π-electron upon oxidation[1,2]. The deformation is named as the electrochemomechanical deformation (ECMD). The ECMDs of conducting polymer, polypyrrole (PPy) [3,4], have exhibited the strain of more than 30 % and the stress of 22 MPa, which are superior to those of skeletal muscles of 25% and 0.5 MPa. Therefore the ECMDs are prospected to be a promising candidate for artificial muscles or soft actuators. We have studied the behaviour of ECMD of PPy films[5,6] under tensile loads up to 5 MPa, and found that the films showed remarkable creeping during electrochemical cycles. The creeping results from breaking, conformation change of polymer backbone and slipping of polymer chains. Upon removal of high tensile stresses, the creeping was recovered to some extent, and interestingly the cyclic stroke became slightly larger than that before the application of the tensile loads. The increased strain of ECMD has been studied as a training effect of the artificial muscles[5,6].

In this paper, ECMD responses of anion driven polyaniline (PANi) actuator operated under various tensile loads are mentioned. Under the relatively large tensile loads, a large creeping in ECMD was observed, and the creeping was recovered upon release of tensile loads. The larger training effect of PANi was obtained compared with the case of PPy. The training (or memory) effect are discussed by the rheology of films under electrochemical cycles taking a model of anisotropic deformation and electrostatic cross linking by oxidation. The large creeping was explained by dynamical flow of ions during electrochemical reaction[7].

EXPERIMENTAL

Emeraldine base (EB) powder of polyaniline (PANi) was synthesized by a chemical method described elsewhere [8,9]. Free-standing PANi films with the thickness of approximately 30μm were prepared by casting a N-methyl-2-pyrrolidinone (NMP) solution of EB on glass substrates. The rectangular films with typical dimensions of 10mm × 2mm × 30μm and weight of 0.1mg were obtained by cutting the film. The film was hung by a platinum clip of working electrode in a special glass cell with a pin hole at the bottom as shown in Fig.1. The other end of film was connected to a thin W wire and a water tank with a reflecting plate through the pin hole. The film was electrochemically cycled to see the actuation in aqueous electrolyte of 1M HCl with Ag/AgCl reference electrode in the cell [8,9]. Various amount of water was put in the tank to give tensile loads. Up and down of the plate due to contraction and expansion of the film, respectively, were measured by a laser displacement meter underneath. The actuator was driven by a computer system with a triangle wave and a rectangular voltage to measure cyclic voltammogram, ECMD and current responses.

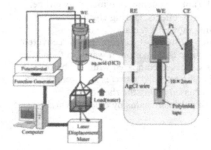

Fig. 1 Experimental setup of in-situ measurement of ECMD of conducting polymer films and the cell construction.

RESULTS AND DISCUSSION

Figure 2 shows the incremental elongation of the ECMD for a PANi film with the original length (L_0) of 10mm operated by a triangle voltage under tensile loads at the scan rate of 2mV/s. The electrochemical cycle was repeated 3 times at the same tensile load as shown by ① ~ ④ in Fig.2. The film showed cyclic strokes of expansion and contraction (ΔL) being 0.4 ~ 0.6 mm (corresponding strain of 4 ~ 6%). The ΔL is induced by insertion and ejection of bulky anions with solvated water due to the electrochemical oxidation and reduction, respectively. ECMD also exhibited a creeping by tensile loads up to 3MPa. The creeping was remarkable for lager tensile loads. It is interesting to note that by release of tensile loads down to 0MPa, the creeping was recovered to some extent as shown in Fig.2. The recovery of creeping suggests that the creeping is partly due to the change of polymer conformation (anisotropic deformation), since breaking and/or slipping of polymer chains results in permanent deformation. It should be also noted that the ΔL as shown by ⑤ and ④ in Fig.2 after 3MPa was substantially larger than ① and ② in Fig.2 for 0 and 1 MPa, respectively, indicating the large training effect.

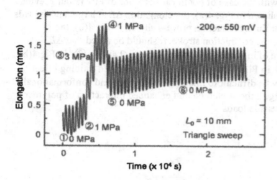

Fig. 2 ECMD of PANi film operated by a triangle sweep at -200 ~ 550mV vs. Ag/AgCl with the period of 900 s under tensile loads up to 3MPa. ①~⑥ indicate the sequence of measurement.

Curves in Fig.3 show the cyclic voltammograms (CV) obtained in PANi film during the electrochemical cycle as shown in Fig.2. By the application of tensile loads, the oxidation peaks shifted to lower potential (easy oxidation) and the reduction peaks also shifted to lower potentials as shown by an arrow. By the removal of tensile load, the CV traced the nearly same curve to that of highest tensile load with a slightly broadening of peaks. The results indicate that some permanent change of polymer conformation occurred.

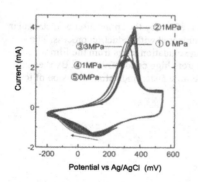

Fig. 3 CV curves during the electrochemical cycles shown by ①~⑥ in Fig.2

Curves in Fig. 4 shows the injected electrical charge (Q) dependences of the elongation (ΔL) in PANi film.The ΔL was obtained at the 2nd cycle of each tensile stress.   The Q was calculated from the current and time from Fig.3. ΔL vs. Q in PANi films showed approximately linear dependence, which can be compared with the case of nonlinear behavior of PPy in our previous report[6]. In the oxidation process as shown in Fig. 4(a), the elongation ΔL at large tensile loads was due to the creeping to some extent. In the reduction process shown in Fig.4(b), the elongation of ΔL during reduction was the contraction stroke. It should be noted that ΔL apparently increased by the experience of high tensile stresses and the Q at ④~⑥ in Fig.4(b) was apparently larger than ①and ② in Fig.4(b), indicating the pronounced training effect. Although the detailed speculation may be difficult to image the electronic and conformational structures, the present results may suggest the stretching and enhanced interaction of polymer chains due the ECMD strain by high tensile loads.

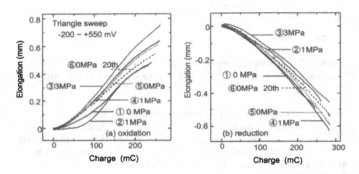

Fig. 4 Elongations of PANi film during (a) oxidation and (b) reduction under tensile loads.

Figure 5 shows the time responses of current waveform by the application of rectangular voltage. It is noted that the oxidation took longer time compared with reduction process. The results are consistent with the case of PPy [6], namely, the oxidation starts from the film with low conductive reduced state, while the reduction starts from high conductive state. By the experience of large tensile loads, the current response became faster, indicating diffusion of ions in the film became faster.

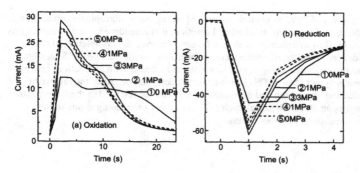

Fig. 5 Time responses of current waveform by the application of rectangular voltage of -200 ~ 550 mV, each 450 s for oxidation and reduction cycles

Table1 indicates the results of ECMD including the energy conversion efficiency in PANi films. The contraction length (stroke, ΔL) of ECMD enhanced significantly by the experience of high tensile load stress (training effect). The contraction stokes of ECMD indicated by the strokes of ④ and ⑤ in Table 1 are much larger than those of ① and ② after the experience of larger tensile loads, namely, the training effect. It is interesting to note that training effect in the PANi film is more pronounced compared with the case of Polypyrrole film [6]. The reason is not clear at the present stage, however, this may be due to the longer conjugation length of PANi polymer chains.

The energy conversion efficiency is estimated by E_m/E_e, where E_e is the electrical input energy, namely, $E_e = \int VI dt$ (V: applied voltage against the reference electrode, I: current) and output work of $E_m = mg\Delta l$ (m: mass of weight, g: the gravity). It is noted that the injected charge (Q) and the input energy were also enhanced by the training. The maximum conversion efficiency was 0.15% at around 3MPa, being similar to 0.15 % in polyaniline actuators by Smela et al. [10]. The small efficiency is due to the charge and discharge processes in the electrochemical reactions [11].

Table 1. Summary of ECMD during the contraction (reduction) in a PANi film under tensile loads.

Tensile Stress (MPa)	Contraction stroke ΔL (μm)	Charge Q (mC)	Input energy E_e (mJ)	Output work E_m (mJ)	Energy conversion efficiency (%)
① 0	490	245	49	0.0048	0.0098
② 1	453	259	51.8	0.0271	0.0523
③ 3	468	282	56.5	0.0843	0.149
④ 1	630	282	56.4	0.0377	0.0668
⑤ 0	635	284	56.7	0.0062	0.0109

The training effect of PPy has been discussed [6] taking the anisotropic deformation and creeping into consideration. During electrochemical reaction, it is considered that polymer chains dynamically fluctuate by flow of anions, and should be easily creped. By completion of the electrochemical oxidation, however, the anisotropic deformation should be fixed or frozen by electrostatic crosslinking as shown in Fig.6(a). The anion may interact with polycations nearby and crosslink adjacent polymer chains. This is a memory of deformation. After the release of tensile stress and reduction, the frozen and expanded state should be released to the original state (recovery of creeping) upon electrochemical cycles. The recovery of creeping is due to the elasticity or thermal vibration of polymer chains.

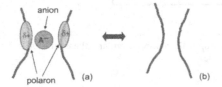

Fig.6 Schematic diagram of (a) creation electrostatic crosslinking by oxidation, and (b) removal of crosslinking by reduction in conducting polymers.

SUMMARY

The ECMD in PANi films has been studied under application of tensile stresses. It was found that the ECMD showed large creeping and the recover of creeping by the removal of tensile loads. The creeping resulted in larger strain of ECMD after release of tensile loads or training effect, which is explained by anisotropic strain and stretch alignment of polymer chains. The enhanced creeping of the film under high tensile stress was discussed taking the rheological of film during electrochemical cycles, where dynamical flow of ions and electrostatic crosslinking (memory state) at oxidized state into consideration.

ACKNOWLEDGEMENTS

This work was supported by a Grant-in-Aid for Scientific Research from the Ministry of Education, Culture, Sports, Science and Technology, Japan.

REFERENCES

1. K. Kaneto, M. Kaneko, Y. Min, Alan G. MacDiarmid, *Synthetic Metals*, 71, Issues 1-3, 1 (1995) pp2211-2212.
2. Y. Sonoda, W. Takashima and K. Kaneto, *Synthetic Metals*, 121 (2001) pp267-268.
3. S. Hara, T. Zama, W. Takashima and K. Kaneto, *Polym. J.*, 36 (2004) p.151.
4. T. Zama, N. Tanaka, W. Takashima and K. Kaneto, *Polymer Journal*, 38 (2006) pp.669.

5. K. Kaneto, H. Suematsu, and K. Yamato, *Adv. Sci. Technol.* 61 (2008) pp122-133.
6. T. Sendai, H. Suematsu, and K. Kaneto, *Jpn. J. Appl. Phys.*, 48 (2009) 051506.
7. J. D.Madden, D. Rinderknecht, P. Anquietil, and I.W. Hunter, *Sensor and Actuators A: Physical*, Vol. 133, Issue 1, (2007) 21-217.
8.   Wataru Takashima, Megumi Nakashima, Shyam S. Pendey, Keiichi Kaneto, *Chemistry Letters*, Vol.32, No.11 (2003) pp990-991
9. Wataru Takashima, Megumi Nakashima, Shyam S. Pandey, and Keiichi Kaneto, *Electrochimica Acta,* Vol.49, Issue 24 (2004) pp.4239-4244.
10. E. Smela, W. Lu and B. R. Mattes, *Synthetic Metals* 151 (2005) p.25.
11. Keiichi Kaneto, Hisashi Fujisue, Masakatsu Kunifusa and Wataru Takashima, *Smart Mater and Structure*, 16 (2007) S250-S255.

Mater. Res. Soc. Symp. Proc. Vol. 1224 © 2010 Materials Research Society   1224-FF06-05-DD06-05

Size-dependent mechanical properties of polymer-nanowires fabricated by two-photon lithography

Satoru Shoji[1], Sana Nakanishi[1], Tomoki Hamano[1], and Satoshi Kawata[1,2],
[1]Department of Applied Physics, Osaka University, Osaka 565-0871, Japan,
[2]Nanophotonics Laboratory, RIKEN, Saitama 351-0198, Japan.

shoji@ap.eng.osaka-u.ac.jp / http://lasie.ap.eng.osaka-u.ac.jp/

ABSTRACT

We present a quantitative characterization of elastic modulus of polymer nanowires. We fabricated a variety of freestanding polymer nanowires in the shape of coil spring with different radii from 100 to 500 nm by means of two-photon lithography. The elastic modulus of each nanowire was measured by applying a mechanical tension onto the nanowire in uniaxial direction by using laser trapping technique. We observed that the elastic modulus and the transition temperature of polymer start to show size-dependent characteristics when the size of the polymer decreases less than 500 nm. We clearly observed that the shear modulus of the polymer nanowires changes from 0.1 MPa to 4.5 MPa when the radius of the nanowire is decreased from 450 nm to 150 nm. Simultaneously, the phase transition temperature of the polymer nanowires also show a significant shift from 35 °C to -10 °C.

INTRODUCTION

Recently polymer nanofibers/nanowires have been getting interest from a variety of academic and industrial fields because polymer possesses many advantageous properties such as biocompatibility, affinity to organic/inorganic materials, mechanical flexibility, elasticity, low toxicity, chemical functionalization flexibility, amorphousness, and so on. For example, polymer nanofibers are one of the candidate materials as a scaffold matrix in tissue engineering. In power engineering, polymer nanofibers have a potential as capacitors for fuel cells. Besides there are many other applications such as chemical membrane filters, clothing application, drug delivery, fiber-reinforced composite materials, nanolithography/nanoprocessing, etc., which have rapidly increased the potential and the importance of nano-polymer materials. Especially, the invention of electrospinning method made us possible to produce nanofibers of a few micrometers or sub-micrometers thickness from many kinds of different polymers [1]. On the other hand, many fundamental properties are not still well understood yet because of the difficulty of manipulating individual sub-micrometer scale thin polymer nanowires to test these properties. Mechanical properties are one of the most important issues to be studied in order to understand the dynamics of polymer chains in nanometer scale, deriving the intrinsic physical/chemical properties of nano-polymers.

Two-photon absorption is now recognized as one of the ideal tool to observe, fabricate, stimulate, and manipulate materials three dimensionally with a spatial resolution beyond the diffraction limit of light. Because of its nonlinear optical response characteristics, two-photon absorption makes it possible to locally initiate various photochemical reactions of materials within a tiny volume of a few hundreds nanometers [2-5]. Since the invention of two-photon lithography [2], two-photon absorption-induced photopolymerization has been widely used as a powerful method to produce three-dimensional polymeric micro/nano structures, such as micro-electromechanical systems (MEMS), micro-fluidic devices, biocompatible micromachines, cell scaffolds, three-dimensional photonic crystals, etc.

In order to achieve reliable functions of the polymer micro/nano devices for practical application, in addition to the spatial resolution or the accuracy of two-photon lithography, it is necessary to understand the physical properties of polymers in nano-scale. The origin of the physical properties of polymers is intra- and inter-molecular dynamics of polymer chains. The spatial resolution of two-photon lithography recently reaches to sub-100 nm and is still being improved by many research efforts. In such a small volume, the dynamics of polymer chains might be quite different from that in the bulk polymer. Indeed, we observed that the elasticity of two-photon lithography-fabricated polymer nanowires shows distinctive features compared to the bulk state polymer [6,7]. We fabricated several freestanding polymer nanowires with different radius, and applied a tensile strain onto the wires by laser trapping technique to study the elasticity of the polymer wires. The results indicate that the mechanical properties of nano-polymers are controllable by the feature size of the polymer.

FABRICATION OF POLYMER NANOWIRES

We prepared a MMA(methyl-methacrylate)-based photopolymerizable resin shown in table 1, consisting of methyl-methacrylate monomer, crosslinker (dipentaerythritol-hexacrylate (DPE-6A)), initiator (benzil), and photosensitizer (IRGACURE 369, Chiba Specialty Chemicals Inc.). The absorption spectrum of the photopolymerizable resin is shown in Figure 1. We utilized two-photon lithography technique to fabricate freestanding polymer nanowires. Figure 2 shows the optical setup of our two-photon lithography system. A light beam of a femtosecond Ti:Sapphire laser at 780 nm was focused by a high numerical aperture (1.4 NA, oil immersion) microscope objective to polymerize the compound at the focus spot. The photosensitizer absorbs the light by two-photon absorption, which initiates photopolymerization reaction at the center of the laser spot. The diameter of the minimum polymerized area (i.e. the spatial resolution of our system) is about 100 nm. A three-dimensional piezo stage driven by a computer scanned the focus spot in the compound to form three-dimensional polymer structures. We fabricated the nanowires in the shape of coil spring (Figure 3). As the geometrical parameters, a coil radius R is 2.5 µm, a number of turns N is 4, and a length of spring L is 8 µm. One end of the spring is attached on an anchor at the height h of 7 µm above the substrate, and a bead with a diameter b of 1 µm is attached on the other end. The nanowire part was fabricated with a single-stroke scan. By controlling the laser power and the scanning speed of the laser spot, we fabricated nanowires with a various diameter r from 100 nm to 500 nm. We prepared a glass container (1 cm wide, 2 mm thick) and put a tiny droplet of the resin (~ 0.1 mm in diameter) on the bottom of the container. We fabricated the springs in the droplet. After scanning the laser beam we poured enough amount of ethanol into the container to rinse un-polymerized resin away. We kept the springs immersed in ethanol, and performed the following elasticity measurement in the solution. Before the measurement, we irradiated UV light to complete the polymerization reaction to guarantee the same polymerization degree in all the nanowires.

Table 1 The MMA-based photopolymerizable resin composition.

	metyl-methacrylate (monomer)	DPE-6A (crosslinker)	benzil (initiator)	IRGACURE 369 (photosensitizer)
Mol (mmol)	6	1		
Weight (%)	50	47	1.5	1.5

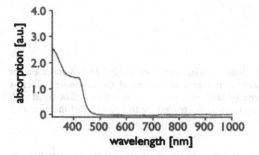

Figure 1 Absorption spectrum of the photopolymerizable resin, showing suitability for two-photon lithography. For one-photon absorption (at 780 nm) the resin has no significant absorption, whereas for two-photon absorption (at 390 nm) the resin possesses high absorption.

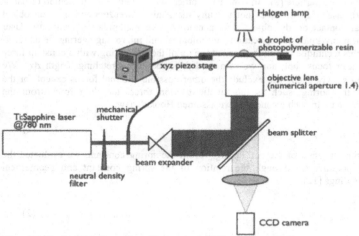

Figure 2 Optical configuration of two-photon lithography system.

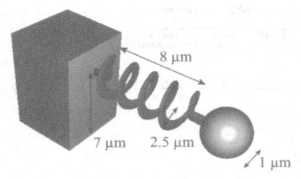

Figure 3 Geometry of a coil spring. The coil-shaped nanowire is suspended from a cubic anchor attached on a glass substrate. A 1-µm bead is fabricated at the end of the nanowire, which is used to apply tensile strain onto the nanowire by laser trapping. The nanowire is fabricated with a single-stroke scan. The radius of the nanowire is controlled by the power and the scanning speed of the laser focus.

MEASUREMENT OF MECHANICAL PROPERTIES OF POLYMER NANOWIRES

For investigating the mechanical properties of the polymer nanowires, we measured the spring constant of each coil spring by applying a mechanical tension through laser trapping technique. Laser trapping is a method to grab a small object by generating an attractive gradient force by laser light on the particle [8]. In contrast to other micro-manipulation method or nano-indentation method, laser trapping does not use any mechanical tweezers and grab an object without mechanical contact on the object. As a non-invasive manipulation technique, laser trapping is used for many kinds of applications in biological and micro-engineering fields [2, 9-11]. We utilized this technique to pull the bead at the tip of the coil spring with a constant force F of 10 pN by a laser focus spot, and we measured the maximum stretching length Δx. We stretched the spring slowly enough to exclude the other possible external forces except for the restoring force of the spring, such as the kinetic friction force, the drag force from the surrounding solution, etc. In such a condition we assumed Hooke's law,

$$F = -k\Delta x,\tag{1}$$

where k is the spring constant of the coil spring. From the spring constant, we evaluated the shear modulus of polymer Gs through the relation of the spring constant and geometrical parameters of coil springs [12],

$$k = \frac{G_s r^4}{4NR_c^{\,3}}\tag{2}$$

We installed a temperature control stage on a laser trapping system so that the temperature of the whole sample was kept stable during the experiment. We used a Nd:YVO4 (Neodymium doped Yttrium Orthvanadate) laser at 1064 nm and a long working distance objective lens (numerical aperture 0.6) to trap the bead. We measured the length of the springs from an optical transmission microscope image through a charge coupled device (CCD) camera. We also performed the same measurement with changing the temperature of the springs to observe the phase transition of polymer nanowires. All of the measurements were performed in ethanol just after the development process.

EXPERIMENTAL RESULTS AND DISCUSSION

Elastic modulus of polymer nanowires and its size dependence

Figure 4(a) is a photograph of two polymer nanowires. The upper one is in the original shape, and the bottom is elongated by being applied a force onto a bead at the end of the wire by means of laser trapping. Figure 4(b) shows the plot of the stretching length Δx versus the applied tensile force for three different nanowires. As shown here, the stretch length is linearly proportional to the force, which guarantees the linear elasticity system and the Hooke's law in a regime of a tensile force from 10 to 60 pN. From the slope of each fitting line, the spring constant was calculated to be 2.60 μN/m (red), 3.46 μN/m (blue), and 3.85 μN/m (yellow). The shear modulus was calculated from the spring constant and the geometrical parameters of the nanowires. Figure 5 shows the shear modulus of the polymer nanowires depending on the radius of the wires. The shear modulus is the measure of elasticity and usually an invariant constant to the size of the materials, whereas our experimental result from the thin wires with a few hundreds of nanometers was not the case. As decreasing the radius of the nanowires, the shear modulus is increasing. The size-dependent behavior becomes obvious when the radius of the nanowires becomes less than about 400 nm. The behavior looks similar to the case of other materials such as metals, semiconductors, etc. In the case of polymers, there are a few reports showing size-dependent elasticity such as polypyrrole nanotubes and polystyrene fibers [13-15]. However, in these cases the materials are highly crystallized, and then the size-dependence in mechanical properties is usually caused by the existence/lack of defects in the crystals, or uniaxial alignment of crystal domains. In contrast, in the case of our experiment polymer nanowires were fabricated by continuous laser-induced photopolymerization during the scan of the laser spot, which in general does not cause any orientation or crystallization of polymer chains. Even though the resultant behavior looks similar, the mechanism of this size-dependence can be quite different.

Another important thing to be noted here is the fact that the measured shear modulus is 3 orders of magnitude smaller than that for the bulk state of the same polymer. The shear modulus of the polymer in bulk state is about 150 MPa. These facts imply that the elastic property of polymer significantly changes when the size of polymer is reduced into sub-micrometer scale. We still do not get a clear conclusion about the reason of this, however one of the possible reasons maybe the penetration of the ethanol into the polymer. Although the affinity of ethanol to PMMA is not high, a small amount of ethanol penetrates into the polymer within the depth of a few tens of nanometers. The penetration of ethanol might lead to unwinding the entanglement of polymer chains and softening the polymer. Since the affinity between ethanol and PMMA is not high such an effect is not seen in the common bulk PMMA. However, when the size of the polymer becomes a few hundreds of nanometers, even the small penetration depth becomes quite a significant quantity to the whole volume enough to affect the elasticity of polymer.

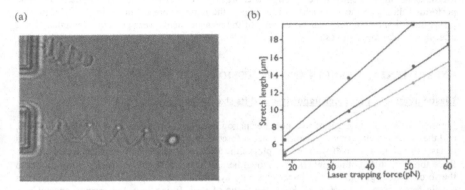

Figure 4 (a) Optical microscope image of polymer nanowire springs fabricated by two-photon lithography. The upper spring is in the original shape, whereas the bottom one is stretched by laser trapping method. (b) Stretching length of the three different nanowires under different laser trapping force. The values of the spring constant are 2.60 μN/m (red), 3.46 μN/m (blue), and 3.85 μN/m (yellow). The stretching length shows linear relation with the pulling force of the laser beams, following the Hooke's law.

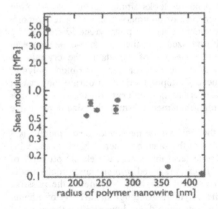

Figure 5 Shear modulus of the polymer nanowires of different radius. Thinner polymer wires show higher shear modulus, although the values of the shear modulus are still 2 ~ 3 orders of magnitude smaller than that of the bulk.

Temperature-dependence on elastic modulus of polymer nanowires

In order to understand the origin of the size-dependent elastic modulus and the remarkable decrease of the stiffness of polymer, we investigated the temperature dependence of the elasticity. Figure 6 is the typical relation of the maximum stretch length on the temperature of polymer coil springs. In the figure, the radii of polymer wire are (a)230 nm, (b)280 nm, and (c)420 nm, respectively. It is clearly seen that the spring constant significantly dropped (i.e. the stretch length increased) when the temperature of the spring reached to certain critical threshold. This result can be interpreted as the glass transition of polymer by which the polymer underwent the transformation from glassy solid state to rubbery solid state.

As seen in Figure 6, we found that the transition temperature in the polymer nanowires also shows size-dependence. A series of the transition temperature (Tc) of different nanowires is plotted in Figure 7. The transition temperature was decreased with the radius of polymer nanowires. Surprisingly, the transition temperature dropped by more than 40 °C with the decrease of the radius from 450 nm to 150 nm. For example, the polymer nanowire of ~ 400 nm in radius is glassy stiff solid, whereas the nanowire of ~ 100 nm shows rubbery elastic property at room temperature. This means that even though we used the same material, same fabrication method, the resultant polymer could be in different amorphous solid state. It must be noted that the polymer in the bulk state shows glass transition at ~ 150 °C and β-transition at 30 °C, both of which were measured by differential scanning calorimetry (DSC).

According to the experimental results, we built our hypothesis about the mechanism of the size-dependent shift of the transition temperature as shown in Figure 8. In microscopic view, polymer consists of three-dimensionally entangled and crosslinked polymer chains forming porous microstructure. At the surface of the polymer, some amount of the solvent surrounding the polymer penetrates into the polymer. The penetration efficiency is related to the affinity between the polymer and the solvent, and the surface of the polymer forms an intermediate matrix of pure polymer and solvent. The penetration of the solution leads to the swell of the polymer, and then the mobility of polymer chains increases. Eventually the increase of the polymer chains' mobility causes the decrease of the transition temperature. The increase of the polymer chains' nobility means softening of polymer, which is consistent to the observation of 2 ~ 3 orders of magnitude smaller elasticity in polymer nanowires compared to the bulk. In order to reveal the evidence of our hypothesis, the possible experiment is to perform the similar mechanical test in other solution. However we haven't successfully done this experiment yet. Because the polymer nanowires were extremely fragile, so that the nanowires were immediately broken by the convection of solution when we added different solvent to exchange ethanol with it into the sample cell.

As shown above, the shear modulus and the transition temperature of PMMA-based polymer are no longer invariable but size-dependent. Similar behavior is also reported in other kinds of polymer from thin films [16,17]. As mentioned, it is known that the mechanical properties of polymers are governed by intra-molecular and inter-molecular interaction of polymer chains. For example, when polymers are heated above their glass transition temperature, the friction between polymer chains starts to become weak, which leads to micro-Brownian motion of the polymer chains. As a result, polymers become softer. The degree of intra- and inter-molecular dynamics in polymer is mainly related to the length of polymer chain and the crosslinking density. Besides the mechanism of solvent penetration mentioned above, there is another possible mechanism. At the surface, polymer chains are more free to move compared to the chains deep inside a bulk of polymer, since loose ends are likely to be abundant and also polymer chains are absolutely free to move in the direction perpendicular to the surface. These kinds of surface effect are usually insignificant where the size of the polymer structure is large enough, however when the size of the structure becomes small the mechanical property of the polymer reflects a large contribution from the surface components. Indeed, the size effect on mechanical properties of polymers is an intrinsic phenomenon appearing in nano-scale dimensions, and so, cannot be neglected when considering micro/nano-device applications.

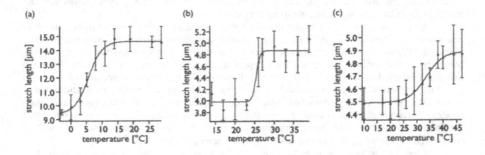

Figure 6   The dependence of the maximum stretch length and the temperature of polymer nanowire coils.   The radii of polymer wire are (a)230 nm, (b)280 nm, and (c)420 nm, respectively. In all the cases, the stretching length increases above certain threshold temperature.

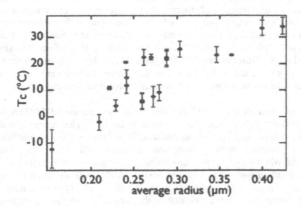

Figure 7 The dependence of the transition temperature (Tc) on the thickness of the polymer wires.

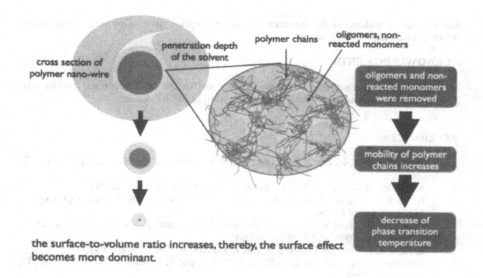

penetration depth of the solvent

polymer chains

oligomers, non-reacted monomers

cross section of polymer nano-wire

oligomers and non-reacted monomers were removed

mobility of polymer chains increases

decrease of phase transition temperature

the surface-to-volume ratio increases, thereby, the surface effect becomes more dominant.

Figure 8 Schematic of the possible mechanism of the nano size effect on the elastic modulus and the phase transition temperature. The critical size threshold where the size effect becomes visible is possibly dependent on the effective penetration depth of the solvent.

CONCLUSIONS

We observed evidence that the mechanical properties of PMMA start to show size-dependence when the size of the polymer becomes in the order of a few hundreds nanometers. We produced a variety of free-standing polymer nanowires with controlled radii by means of two-photon lithography. We performed quantitative analysis of the mechanical properties for each polymer nanowire in ethanol using laser trapping technique. We clearly observed that the shear modulus of the polymer nanowires changes from 0.1 MPa to 4.5 MPa when the radius of the nanowire is decreased from 450 nm to 150 nm. Simultaneously, the phase transition temperature of the polymer nanowires also show a significant shift from 35 °C to -10 °C. The increase of the elastic modulus and the decrease of the phase transition temperature occurring at the same time seem to be contradictory each other, and are not easy to be understood. The mechanism behind the observed size-dependent behavior in the elasticity and the phase transition temperature is still not clear yet. The mechanical test of the polymer nanowires in dried condition would give us another clue to understand the dynamics of supermolecular system of polymer in nanoscale, which is a next step of this research.

ACKNOWLEDGMENTS

This research is supported by Iketani Science and Technology Foundation, and Grant-in-Aid for Young Scientists (no.21686010, and no.19810012), MEXT, Japan.

REFERENCES

1.   Z.-M. Huang, Y.-Z. Zhang, M. Kotaki, and S. Ramakrishna, Compos. Sci. Tech. **63**, 2223 (2003).
2.   S. Kawata, H.-B. Sun, T. Tanaka, and K. Takada, Nature, **412**, 697 (2001).
3.   H. Ishitobi, S. Shoji, T. Hiramatsu, H.-B. Sun, Z. Sekkat, and S. Kawata, Opt. Express, **16**, 14106-14114 (2008).
4.   Q. Ya, W.-Q. Chen, X.-Z. Dong, T. Rodgers, S. Nakanishi, S. Shoji, X.-M. Duan, and S. Kawata, Appl. Phys. A, **93**, 393-398(2008).
5.   S. Kawata, S. Shoji, and H.-B. Sun, IEICE T. Electron., **E87C**, 378(2004).
6.   S. Nakanishi, H. Yoshikawa, S. Shoji, Z. Sekkat, and S. Kawata, J. Phys. Chem. B **112**, 3586-3589 (2008).
7.   S. Nakanishi, S. Shoji, S. Kawata, and H.-B. Sun, Appl. Phys. Lett. **91**, 063112 (2007).
8.   A. Ashkin, J. M. Diedzic, F. E. Bjorkholm, and S. Chu, Opt. Lett. **11**, 288(1986).
9.   A. Ashkin, J. M. Dziedzic, and T. Yamane, NATURE, **330**, 769(1987).
10.   A. Ashkin and J. M. Dziedzic, Science, **235**, 1517(1987).
11.   Y. Inouye, Satoru Shoji, H. Furukawa, O. Nakamura, and S. Kawata, Jpn. J. Appl. Phys., **37**, L684, June (1998).
12.   C. J. Ancker and J. N. Goodier, J. Appl. Mech. 25, 466 (1958).
13.   A. Arinstein, M. Burman, O. Gendelman, and E. Zussman, Nature Nanotechnol. **2**, 59 (2007).
14.   S. Cuenot, S. Demoustier-Champagne, and B. Nysten, Phys. Rev. Lett., **85**, 1690 (2000).
15.   Y. Ji, B. Li, S. Ge, J. C. Sokolov, and M. H. Rafailovich, Langmuir, **22**, 1321 (2006).
16.   J. A. Forrest and K. Dalnoki-Veress, Adv. Colloid Interface Sci. **94**, 167 (2001).
17.   M. Alcoutlabi and G. B. McKenna, J. Phys.: Condens. Matter. **17**, R461 (2003).

Mater. Res. Soc. Symp. Proc. Vol. 1224 © 2010 Materials Research Society   1224-FF10-23

Electrospun Polymer/MWCNTs Nanofiber Reinforced Composites "Improvement of Interfacial Bonding by Surface Modified Nanofibers"

Elif Ozden, Yusuf Menceloglu, Melih Papila
Sabanci University, Material Science and Engineering, 34956, Tuzla, Istanbul, Turkey

ABSTRACT

In-house synthesized copolymers Polystyrene-co-glycidyl methacrylate (PSt-co-GMA) are electrospun as mat of surface modified nanofibers with and without multi walled carbon nanotubes (MWCNTs). Composites are then formed by embedding layers of the nanofiber mats into epoxy resin. Interfacial bonding between polymer matrix and the nanofibers, and surface modification driven enhancement in mechanical response is assessed under flexural loads. Results indicate that at elevated temperature storage modulus of epoxy reinforced by PSt-co-GMA nanofibers and PSt-co-GMA/ MWCNTs composite nanofibers is about 10 and 20 times higher than the neat epoxy, respectively, despite weight fraction of the nanofibers being as low as 2 wt%. Interfacial interaction is revealed by the storage modulus comparison of unmodified Polystyrene (PSt) and modified PSt-co-GMA nanofiber reinforced composite. To enhance further the resulting "crosslinked" structure, crosslinking agent ethylenediamine is also sprayed on the nanofibrous mats. Increased crosslinking density improves mechanical response of sprayed-over PSt-co-GMA nanofibers reinforced composites which is about 4 times higher than plain PSt-co-GMA nanofibers.

INTRODUCTION

Nano-scaled materials in composites are in increasing demand due to their significant potential for improved materials properties. In different forms and chemistry nano-scaled fillers, such as carbon nanotubes, organic and inorganic nanofibers offer variety of applications and performance enhancements in the field of composites, adhesives and coatings [1-2]. Tailoring the properties of composites depends strongly on the filler size, shape, distribution degree, the filler-filler and filler-matrix interactions [3-5].

Carbon nanotubes (CNTs) have been widely considered as a filler material due to their unique mechanical and electrical properties such as high specific strength and stiffness and electrical conductivity, [6-8]. Nano- to submicron-scale fibers are also recently explored for their reinforcing ability in composites. To increase the matrix-dominated flexural properties for instance electrospun nanofibers were introduced as reinforcement [9]. In addition, there are numerous attempts to fabricate CNTs added electrospun poymer nanofiber webs, to improve mechanical properties of the polymer matrix. [3,10-12]. Moreover, strong surface interactions enable good mechanical interlocking with surrounding polymer chains [13] and upgrade the physical properties of nanocomposites.

In this study, the hypothesis is that the incorporation of electrospun surface modified nanofibers which have functional groups as epoxide ring into the epoxy resin results in significant improvements of the mechanical properties of the polymer matrix. Polystyrene-co-glycidyl methacrylate P (St-co-GMA) and PSt-co-GMA with MWCNTs composite nanofibers were produced by electrospinning. Due to its chemical nature, PSt-co-GMA is expected to improve interfacial bonding with the epoxy based polymer matrix. The problem of MWCNTs dispersion in the polymer matrix was also addressed by introducing them via electrospinning and resultant PSt-co-GMA/MWCNT composite nanofibers. In addition to reinforcement by the presence of nanofibrous layers, the experimental procedure was designed to see the effects of the GMA composition in the structure and the effect of additional crosslinking agent (ethylene diamine) by spraying method. These multiple factors, chemistry or functional groups of the nanofibers, dispersed MWCNT and crosslinking agent, on the mechanical response of the epoxy based composite reinforced nanofibers has been investigated by dynamic mechanical analysis (DMA) instrument. The effects of PSt-co-GMA and PSt-co-GMA /MWCNTs nanofibers on the resulting composites thermal stability were obtained by differential scanning calorimetry (DSC). Universal testing machine was also utilized to determine flexural modulus and flexural strength of neat epoxy and reinforced composite.

EXPERIMENTAL

Copolymers Polystyrene-co-glycidylmethacrylate (PSt-co-GMA) which consist 10 % wt GMA amount were synthesized by solution polymerization. Polystyrene was synthesized also by using the same polymerization technique. Polymer solutions (PSt-co-GMA/ DMF), PSt/DMF and (PSt-co-GMA/ DMF with MWCNTs) at %30 wt concentration were prepared at room temperature. The solutions were stirred magnetically for 24 hour to obtain homogeneity. Electric charge (via Gamma High Voltage ES 30P-20W) was applied to the polymer solutions contained in 2 ml syringe which has an alligator clip attached to the syringe needle (diameter 300 μm). Applied voltage was adjusted to 15kV while the grounded collector was placed 10 cm away from the syringe needle to form the electropinning set-up. Syringe pump (NewEra NE-1000 Syringe Pump) was utilized to control the solution flow rate which was 30 μl/hr during electrospinning. The collected electrospun nanofiber mats were cut into pieces with the size of 12 mm x 50 mm for DMA test. These nonwoven fibrous webs were then embedded into epoxy matrix to form the polymer composites. Custom designed teflon mold was used to produce 2 mm x 12 mm x 50 mm beam shape composite specimens. To remove the trapped bubbles, vacuum was applied. Epoxy reinforced with 2 wt% mass fraction of nanofibers was cured at 50° for 15 hours.

The composition of synthesized P(St-co-GMA) copolymers were determined by C-NMR and H-NMR. All the H-NMR and C-NMR experiments were performed in CDCl3 on Varian Unity Inova 500MHz spectrometer. Glass transition temperatures (Tg) of nanofiber reinforced composites were obtained using Netzsch DSC- 204 instrument. The morphologies of PSt-co-GMA and PSt-co-GMA/MWCNTs fibrous webs and fracture surfaces of neat epoxy and reinforced composites were evaluated by scanning electron microscope (SEM - LEO 1530VP). By using a Rotational Rheometer (Malvern BOHLIN CVO), the shear viscosity of solutions for electrospinning process was also measured at control shear stress of 10 Pa to 1000 Pa. Dynamic thermal-mechanical properties of neat epoxy and nanofiber reinforced composites were investigated by using a dynamic mechanical thermal analyzer (Netzsch DMA 242). The mechanical tests of ten samples for each specimen were performed in three-bending mode at a frequency of 1 Hz within 20°- 150° C temperature range. Maximum amplitude, maximum dynamic force and static constant force were determined as 30 μm, 3 N and 0,01 N, respectively. Universal testing machine (UTM) was finally employed to determine flexural strength and flexural modulus using ASTMD 790 standards, tests were repeated five times for each specimen.

DISCUSSION

Electrospinning of P(St-co-GMA) , P(St-co-GMA)/MWCNTs and PSt nanofibers

Three different polymer solutions, P(St-co-GMA), P(St-co-GMA)/MWCNTs and PSt were electrospun under the same processing conditions. Styrene monomer was chosen for its aromatic ring, which should interact favorably with CNTs. [14] Long term stabilization of MWCNTs in electrospinning polymer solution has been successfully achieved during nanofiber formation. To prove that argument, Dynamic Light Scattering (DLS) was employed, same amount of MWCNTs in DMF and P(St-co-GMA)/ DMF solution were compared. Average particle size of MWCNTs/ DMF solution was 800 nm. On the other hand, two peaks were observed in P(St-co-GMA)/DMF solution. These peaks were around 10 nm and 500 nm, with average MWCNTs particle size around 330 nm. There was no precipitation in the electrospinning solution. SEM images demonstrated that the diameter of PSt and P(St-co-GMA) nanofibers was in the range of 400 – 800 nm, while the diameter range of P(St-co-GMA)/MWCNTs nanofibers was 200 - 550 nm. That is, P(St-co-GMA)/MWCNTs nanofibers were considerably thinner than PSt and P(St-co-GMA) nanofibers. This is attributed to the shear thinning effect by the MWCNTs. In order to validate, shear viscosity of the polymer solutions were measured by rotational rheometer. Results indicated that addition of MWCNTs in PSt-co-GMA/DMF solution reduced shear viscosity under controlled shear stress (see Figure 1). The viscosity of the polymer solution was one of the impact factors on the electrospun nanofiber diameter. Due to shear thinning behavior of MWCNTs in polymer solution, shear viscosity decreased and resulted in reduced fiber diameter. It is worthy to note that our earlier studies pointed that fiber diameter tended to increase with polymer concentration and decrease with MWCNTs concentration.

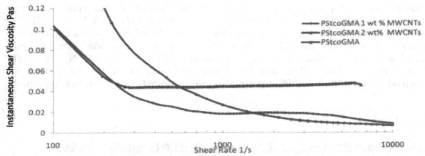

Figure 1. Instantaneous Shear Viscosity vs Shear Rate measure of DMF and polymer solutions at the same concentration. Shear thinning effect of MWCNTs could be observed.

Glass Transition Temperature and Damping Ratio of Reinforced and Unreinforced Epoxy

Glass transition temperatures of nanofiber reinforced composites and neat epoxy were obtained both from DMA and DSC. Eventhough, Tg by DMA may differ from Tg by DSC, they have the same tendency per factors evaluated. For instance, Tg-DMA curve was shifted to a higher temperature in the case of PSt and PSt-co-GMA as in the Tg curve in DSC. On the other hand, Tg of PSt-co-GMA/ MWCNTs nanofiber reinforced epoxy was lowered. Additives as in the case of MWCNTs decrease the Tg, which is also related to viscosity and curing process.

Tanδ results from DMA are considered as damping ratio. The damping ratio of composite materials reflects the ability of the materials to dissipate energy, which mainly comes from the interaction among different inner phases [15]. In the case of strong interfacial bonding between polymer matrix and nanofibers, the energy dissipation can be expected to reduce and the damping ratio should be decreased. As anticipated, the damping ratio of nanofiber reinforced composites suggested the strong interaction and low damping-high modulus relation. For instance, damping ratio of PSt nanofiber reinforced composites were higher than strongly bonded PSt-co-GMA nanofiber reinforced composites (Figure 2). Damping ratio comparison is considered herein as another pointer to explain interfacial bonding between nanofiber and polymer matrix.

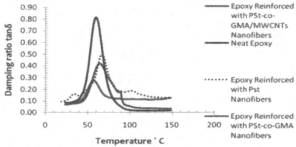

Figure 2. Damping ratio tanδ vs temperature of reinforced and unreinforced epoxy specimens

Dynamic Mechanical Analysis and UTM Mechanical Tests

The experiments were designed in order to investigate several factors including reinforcement with PSt-co-GMA and PSt-co-GMA/MWCNTs nanofibers, additional crosslinking agent ethylene diamine spraying and GMA structure in the interface. Results indicated that at elevated temperature storage modulus of epoxy reinforced by PSt-co-GMA nanofibers and PSt-co-GMA/ MWCNTs nanofibers (both at 2 wt% mass fraction of the fibrous reinforcements (10 layers)) was about 10 and 20 times higher than the neat epoxy, respectively. Interfacial interaction was also revealed by the storage modulus comparison of unmodified and modified nanofiber (PSt and PSt-co-GMA, respectively) reinforced composites. Reinforcement of plain PSt nanofibers could not form strong covalent bonds with epoxy matrix as suggested by the observation that 2% wt PSt nanofiber reinforced composite did not exhibit significant enhancement compared to neat epoxy particularly at elevated temperatures. On the other hand, it was observed that due to consisting GMA composition, PSt-co-GMA nanofiber reinforced composites have considerable increase on storage modulus (Figure 3). Moreover, the resulting "crosslinked" structure was further enhanced when crosslinking agent ethylenediamine was sprayed on the nanofibrous mats. Increased crosslinking density improved mechanical response of crosslinker sprayed-over PSt-co-GMA nanofibers reinforced composites which is about 4 times higher than composites of the plain PSt-co-GMA nanofiber reinforcements.

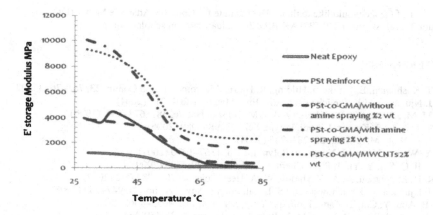

Figure 3. E' storage modulus vs Temperature curves of PSt, PSt-co-GMA with/out modification, PSt-co-GMA/ MWCNTs and neat epoxy.

ASTM- D790 3 point bending standard mechanical tests also demonstrated that embedding a single layer of PSt-co-GMA nanofibrous mat and a *single layer* PSt-co-GMA/ MWCNTs nanofibrous mat was altered flexural modulus about 7% and 22%, correspondingly. Flexural strength of neat epoxy, single layer PSt-co-GMA nanofiber reinforced and single layer PSt-co-GMA/ MWCNTs nanofiber reinforced were 89 MPa, 94 MPa, 100 MPa, correspondingly.

CONCLUSIONS

The effect of interfacial bonding enhanced by surface modification on reinforcing fibrous mat was studied and proved by interpreting the DMA results. Electrospinning technique was used to fabricate surface modified PSt-co-GMA and PSt-co-GMA/MWCNTs nanofiber in order to reinforce the epoxy resin. To our knowledge, despite several efforts towards electrospun nanofiber reinforced composites, this study is the first to report how the chemistry, surface modification and crosslinking agent when associated with the choice of polymer matrix enhance the mechanical behavior of such composites via improved interfacial bonding. Storage modulus of epoxy reinforced by PSt-co-GMA nanofibers and PSt-co-GMA/ MWCNTs composite nanofibers is about 10 and 20 times higher than the neat epoxy, respectively, at elevated temperature despite weight fraction of the nanofibers being as low as 2 wt%. Addition of croslinking agent by spraying-over the reinforcement resulted in further improvement in the mechanical response.

ACKNOWLEDGMENTS

Elif Özden would like to thank The Institute for Complex Adaptive Matter (I2CAM), for Junior Travel Award and TUBITAK BIDEB graduate student scholarship

REFERENCES

1. T. Kashiwagi, E. Grulke, J. Hilding, R. Harris, Macromol. Rapid Comm. 23, 761–765 (2002).
2. J. Njuguna and K. Pielichowski, Adv. Eng. Mater., 6, 204–210 (2004).
3. M. M. J Treacy, T. W Ebbesen and J. M. Gibson, Nature, 381, 678–680. (1996).
4. T. Uchida, S. Kumar, J. Appl. Polym. Sci., 98, 985–989 (2005).
5. D.H. Walt, MRS Bull., 29, 281–285 (2004).
6. Y. Wan, J. He and J. Yong Yu, Polym Int 56, 1367–1370 (2007).
7. J. Ji, G. Sui, Y. Yu, J. Phys. Chem. *C*, 113 (12), 4779-4785 (2009).
8. R. H. Baughman, A. A. Zakhidov, W. A. Heer, Science, 297, 787-792 (2002)
9. J. Njuguna, K. Pielichowski and S. Desai, Polym. Adv. Technol.19: 947–959 (2008)
10. B. Ahn, Y. Chi, T. Kang, J. of App. Poly. Sci.,Vol. 110, 4055–4063 (2008)
11. R. Sen, B. Zhao, D. Perea, M. E. Itkis, *Nano Letters*, 4 (3), 459-464 (2004)
12. C. Seoul, Y. Kim, C. Baek, J. Poly. Sci. Part B: Poly. Phys., Vol. 41, 1572–1577 (2003)
13. J. Suhr, N.A. Koratkar, D. Ye and T. Lu, J. Intel. Mat. Sys.and Struct., Vol. 17 (2006)
14. M.R. Nyden, S.I. Stoliarov, Polymer 49, 635-641(2008)
15. L. Chen, X.L. Gong and W.H. Li, Polymer Testing Vol. 27, 3, Pages 340-345 (2008)

Mater. Res. Soc. Symp. Proc. Vol. 1224 © 2010 Materials Research Society   1224-FF10-19

Multiscale Computer Simulation of Tensile and Compressive Strain in Polymer-Coated Silica Aerogels.

Brian Good
Materials and Structures Division, NASA Glenn Research Center, Cleveland, Ohio.

ABSTRACT

While the low thermal conductivities of silica aerogels have made them of interest to the aerospace community as lightweight thermal insulation, the application of conformal polymer coatings to these gels increases their strength significantly, making them potentially useful as structural materials as well. In this work we perform multiscale computer simulations to investigate the tensile and compressive strain behavior of silica and polymer-coated silica aerogels.

Aerogels are made up of clusters of interconnected particles of amorphous silica of less than bulk density. We simulate gel nanostructure using a Diffusion Limited Cluster Aggregation (DLCA) procedure, which produces aggregates that exhibit fractal dimensions similar to those observed in real aerogels. We have previously found that model gels obtained via DLCA exhibited stress-strain curves characteristic of the experimentally observed brittle failure. However, the strain energetics near the expected point of failure were not consistent with such failure. This shortcoming may be due to the fact that the DLCA process produces model gels that are lacking in closed-loop substructures, compared with real gels. Our model gels therefore contain an excess of dangling strands, which tend to unravel under tensile strain, producing non-brittle failure. To address this problem, we have incorporated a modification to the DLCA algorithm that specifically produces closed loops in the model gels.

We obtain the strain energetics of interparticle connections via atomistic molecular statics, and abstract the collective energy of the atomic bonds into a Morse potential scaled to describe gel particle interactions. Polymer coatings are similarly described. We apply repeated small uniaxial strains to DLCA clusters, and allow relaxation of the center eighty percent of the cluster between strains. The simulations produce energetics and stress-strain curves for looped and nonlooped clusters, for a variety of densities and interaction parameters.

INTRODUCTION

Silica aerogels are low-density, highly porous materials possessing thermal properties that have made them of interest for a wide variety of applications [1-3]. Notably, the low thermal conductivities characteristic of such gels have led to the aerospace community's interest in these materials as lightweight thermal insulation. While these aerogels' fragility limits their utility in many applications, researchers in our laboratory have developed a method for monomer-coating aerogels and cross-linking the coatings, so as to greatly improve the gels' strength while not greatly impacting their insulating properties [4]. Such coated gels may prove suitable for use as lightweight structural materials.

In order to provide an understanding of the mechanical behavior of the gels, and to provide predictive tools of use in their further development, we have constructed a multiscale model for the tensile and compressive failure of pristine and polymer-coated silica aerogels. The model is built on computer simulations using a modified diffusion-limited cluster aggregation (DLCA) scheme [5], along with a particle-based molecular statics procedure.

Earlier work using this model at times produced tensile failure behavior that was less brittle than what is experimentally observed. Visualization of these results suggests that there may be an unphysical uncoiling of the strands of secondary particles produced by the DLCA method, giving strain behavior more ductile than is observed experimentally. To correct this deficiency, we have added an additional step to the structural model generation process, in which strands having one free end are encouraged to attach themselves to the gel cluster that is the final result of the DLCA process, resulting in a more looped (and possibly more brittle) structure.

We perform a variety of multiscale computer simulations to investigate the tensile and compressive strain behavior of silica and polymer-coated silica aerogels. We discuss the strain behavior via energetics and stress-strain curves, for both tensile and compressive strain, for both looped and unlooped DLCA clusters.

STRUCTURAL MODEL

Experimentally, aerogels appear to consist of disordered aggregates of connected fractal clusters, with fractal behavior evident over a limited range of length scales [6-9]. In more detail, aerogels exhibit a low-density "pearl-necklace" structure that consists of tangled strands of approximately spherical particles. These "secondary" particles exhibit internal structure, consisting of smaller "primary" particles of amorphous silica of less than bulk density; an in-chain density of about 1.8 g/cm^3 has been reported by Woignier *et al.* [10]. Two examples of silica gels are shown in Figure 1.

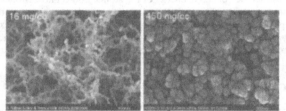

Figure 1. Silica Aerogels. 0.016 g/cm^3 (left), 0.450 g/cm^3 (right).

Model gel structures are produced using the Diffusion Limited Cluster Aggregation (DLCA) scheme. During aggregation, a computational cell is seeded with randomly-distributed secondary particles having a uniform radius of 15 nm. Particles are moved at random, forming subclusters, and, eventually, one large cluster. More specifically, a particle or subcluster is chosen at random, with a probability given by $P_i = (m_i/m_0)^a$, where m_i and m_0 are the masses of the chosen subcluster and the lightest subcluster remaining in the computational cell, respectively, and a is a scaling parameter, here taken to be -0.5. The subcluster is moved in a random direction, and is then inspected for collisions with other particles or subclusters; if one or more collisions occur, the colliding subclusters and particles are merged into a single subcluster. Aggregation proceeds until only a single cluster remains.

In addition, to investigate the effects of looped structures in the DLCA clusters we have added a post-DLCA stage to cluster generation. All dangling strands are identified; each strand is traced from its end (a particle having a coordination of one) back to a "pivot point," a particle having coordination of three or greater. The "target" particle nearest to, but outside the strand is identified, and if possible the strand is rotated so as to allow the strand end to bond to the target particle, forming a loop. Because the looping process takes place after the DLCA process is complete, it is possible to compare the behavior of looped and unlooped versions of the same computational cell. We further assume that, in the case of polymer-coated gels, that the coatings bond conformally to the silica gel and do not significantly change the topology.

It should be noted that the looping process described above is different from the DLCA/Dangling Bond Deflection (DLCA-DEF) model of Ma et al. [11], in which loop formation takes place during the DLCA process.

MULTISCALE FAILURE MODEL

The clusters obtained from the structural model consist of spherical particles in contact. Micrographs suggest that real gel secondary particles are connected by interparticle "bridges" whose diameters are smaller than those of the connected particles. Polymer coatings are assumed to be of uniform thickness over particles and bridges, and failure is assumed to occur only through the bridges.

Gel failure simulations are performed using a multiscale technique, in which the strain energetics of the interparticle bridges and polymer coatings are described by simple potentials whose parameters are obtained from higher-fidelity atomistic simulations. A cylindrical atomistic bridge model of amorphous silica is given repeated small axial strains, with atomistic relaxation between strains accomplished via molecular statics. Atomistic interactions are computed using a Morse potential. The resulting energy-versus-strain curves are suggestive of the Universal Binding Energy Relation (UBER) of Rose, Smith and Ferrante [12], and we fit the bridge energy versus strain curve to a larger-scale interparticle Morse potential, as this form, with appropriate scaling, is known to accurately represent the UBER.

Strain energetics of the polymer coatings are treated in a similar manner. Because several polymer coatings are under development, we have chosen to use generic potential parameters for the Morse polymer potentials, characterizing them with respect to the silica gel interparticle potential parameters.

RESULTS AND DISCUSSION

We have performed simulations for three densities of uncoated gels, and for coated gels derived from the same models, for a range of coating strengths, in both looped and unlooped states. We characterize the uncoated densities as low (0.029 g/cm^3), medium (0.067 g/cm^3) and high (0.184 g/cm^3). The silica interparticle potential well depth parameter V_0 is 2.5-e10 ergs and the well depths for the coatings are 1.0e-12 ergs (weak), 2.5e-10 ergs (medium) and 2.5e-9 ergs (strong). The coating thickness is assumed to be 0.1, normalized to the radius of the interparticle silica bridge. With secondary particle interactions described by the large-scale Morse potential as above, a tensile strain is applied to each DLCA cluster, using a molecular statics procedure, until the energy reaches a maximum (tensile strain) or shows an abrupt increase (compressive strain).

Representative plots of energy versus strain for tensile and compressive strain are shown in Figure 2. Energies are measured with respect to the energy of the corresponding relaxed but unstrained cell. The tensile energies exhibit qualitatively similar behavior—the curvature is positive at small strain, but become negative at larger strain. In previous work we have found that the energy under tensile strain reaches a broad maximum at larger strains. Compressive energies, on the other hand, show only positive curvature, with no indication of failure at the strains studied. It is known that, because of the high degree of porosity, aerogels under compression may exhibit densification before failure,

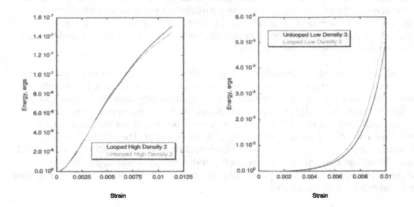

Figure 2. Tensile strain energetics, high density (a); Compressive strain energetics, low density (b).

and that is consistent with the energetics seen here, although experimentally the densification occurs over a broader range of compressive strain. It is evident that compressive failure occurs at a larger strain than is shown here. It should be noted that all plots shown use data taken from weak-polymer simulations. Because the forms of the silica and polymer interparticle potentials are similar, no qualitative differences are observed among simulations using different polymer potential strengths.

It is apparent that, while the looped version of a model gel cluster typically exhibits higher energy and stress at a given strain, the shapes of the energy and stress-strain curves for looped and unlooped cells are very similar. This suggests that adding a looping stage to model gel formation does not qualitatively change the failure mechanism. Ma *et al.* have found that when the looping process is carried out during the DLCA process, rather than afterward, model gel behavior is more brittle.

It can be seen that the variation in energy among different clusters having the same density but obtained from different randomized initial states of the computational cells before aggregation is comparable in magnitude to the difference between looped and unlooped versions of the same cell. However, it should be noted that in almost all cases, the looped cell exhibits a larger energy, and higher stress, at a given strain.

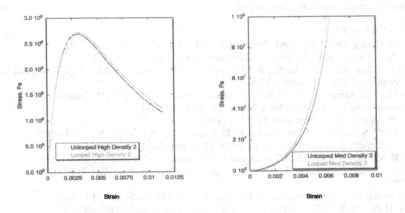

Figure 3. Stress versus strain. Tensile high density (a); Compressive medium density (b).

Stress-strain curves are shown in Figure 3. Under small tensile strain, the model gels exhibit behavior consistent with brittle failure; the stress increases sublinearly before reaching a maximum. There is no abrupt decrease in the stress indicative of sudden failure, but this may be attributed to uncoiling of the strands within the cluster. If we define the point of failure as the strain at which maximum stress is reached, our values of maximum stress are consistent with the work of Meador et al. [13], who report failure in the range of a few MPa. Our strain at failure values, however, are approximately 0.3 percent, considerably smaller than typical experimental values of about a percent. As is the case with the energies, the stress at a given strain is slightly larger for the looped version of the cells than for the unlooped ones. Under compressive strain, the stress increases without exhibiting a maximum, consistent with densification below the strain at failure.

CONCLUSIONS

We have performed multiscale computer simulations of the tensile and compressive failure of polymer-coated silica aerogels. The simulations of gel structure were based on a diffusion-limited cluster aggregation procedure, including, in some cases, an additional looping stage. Strain energetics and stress-strain behavior were modeled using a molecular statics procedure. Simulations were carried out at densities representative of real aerogels. Tensile stress-strain curves exhibit characteristics suggestive of brittle failure. Model gel stresses are consistent with experiment, but the strain at failure is smaller than observed experimentally. Compressive strain produced behavior characteristic of densification, but no compressive failure at the strain values used in the simulations. Looped cells exhibit slightly larger energies and stresses than do the corresponding unlooped ones, but the energy and stress-strain curves are very similar in shape, indicating that adding loops to the DLCA clusters after cluster formation does not fundamentally change the behavior of the model.

REFERENCES
1. N. Husing and U. Schubert, Angew. Chem. Int. Ed. **37**, 22 (1998).
2. A. C. Pierre and G. M. Pajonk, Chem. Rev. 2002, **102**, 4243 (2002).
3. J. L. Rousset, A. Boukenter, B. Champagnon, J. Dumas, E. Duval, J, F, Quinson and J. Serughetti, J. Phys.: Condens. Matter **2**, 8445 (1990).
4. M. A. B. Meador, E. F. Fabrizio, F. Ilhan, A. Dass, G. Zhang, P. Vassilaras, J. C. Johnston and N. Leventis, Chem. Mater. **17**, 1085 (2005).
5. P. Meakin, Phys. Rev. Lett. **51**, 1119 (1983).
6. D. W. Schaeffer, J. E. Martin and K. D. Keefer, Phys. Rev. Lett. **56**, 2199 (1986).
7. T. Freltoft, J. K. Kjems and S. K. Sinha, Phys. Rev. B **33**, 269 (1986).
8. R. Vacher, T Woignier, J. Pelous and E. Courtens, Phys. Rev. B **37**, 6500 (1988).
9. A. Hasmy, E. Anglaret, M. Foret, J. Pelous and R. Julien, Phys. Rev. B **50**, 1305 (1994).
10. Woignier : T. Woignier and J. Phalippou, J. Non-Cryst. Solids **93** 17 (1987).
11. H.-S. Ma, R. Jullien and G. W. Scherer, Phys. Rev. E **65**, 041403 (2002).
12. J. H. Rose, J. Ferrante and J. R. Smith, Phys. Rev. Lett. 47, 675 (1981).
13. M.A.B. Meador, S. L. Vivod, L McCorkle, D. Quade, R. M. Sullivan, L. J. Ghosn, N. Clark and L. A. Capadona, J. Mater. Chem. **18**, 1852 (2008).

Mater. Res. Soc. Symp. Proc. Vol. 1224 © 2010 Materials Research Society

Correlating Nanoparticle Dispersion to Surface Mechanical Properties of TiO₂/Polymer Composites

Yongyan Pang, Stephanie S. Watson, Aaron M. Forster, and Li-Piin Sung

Polymeric Materials Group, Building and Fire Research Laboratory, National Institute of Standards and Technology, Gaithersburg, MD 20899

ABSTRACT

The objective of this study is to characterize the nanoparticle dispersion and to investigate its effect on the surface mechanical properties of nanoparticle-polymer systems. Two types of TiO₂ nanoparticles were chosen to mix in two polymeric matrices: solvent-borne acrylic urethane (AU) and water-borne butyl-acrylic styrene latex (latex) coatings. Nanoparticle dispersion was characterized using laser scanning confocal microscopy. Overall, Particle A (P_A, without surface treatment) dispersed better than Particle B (P_B, organic treatment) in both systems. The AU-P_A system exhibited the best dispersion of the four systems, however P_B forms big clusters in both of the matrices. Surface mechanical properties, such as surface modulus at micron and sub-micron length scales were determined from depth sensing indentation equipped with a pyramidal tip or a spherical tip. The surface mechanical properties were strongly affected by the dispersion of nanoparticle clusters, and a good correlation was found between dispersion of nanoparticle clusters near surface and the modulus-depth mapping using a pyramidal tip.

INTRODUCTION

The addition of nanoparticles into polymeric coatings and composites has potential to improve performance and increase the number of applications compared to their micron-sized pigmentary counterparts [1-3]. However, poor dispersion and agglomeration of nanoparticles in polymeric matrices still remain a challenge in research and industrial applications. Agglomeration and poor dispersion adversely affect appearance, service life, and mechanical properties of polymer nanocomposites. Traditional mechanical testing methods, such as dynamic mechanical thermal analysis and tensile tests, detect changes in bulk mechanical properties upon the addition of nanoparticles; however, they are not always sensitive to the changes in local structural features. In this study, laser scanning confocal microscopy (LSCM) was used to map TiO₂ nanoparticle dispersion in polymeric coatings and depth sensing indentation (DSI) techniques were used to measure the film mechanical properties within the first few micrometers from the surface. Nanoparticle-polymer coating systems having different dispersion states were prepared with two types of TiO₂ nanoparticles incorporated into two different polymeric matrices. The effect of indenter acuity on the material indentation response was investigated using pyramidal and spherical indenters. The impact of particle dispersion on local modulus was investigated by imaging the area after indentation. Good correlation was found between nanoparticle dispersion and surface modulus mapping using a pyramidal tip.

EXPERIMENTAL[#]
Materials

[#] Certain instruments or materials are identified in this paper in order to adequately specify experimental details. In no case does it imply endorsement by NIST or imply that it is necessarily the best product for the experimental procedure.

Two commercially available nanoparticles were selected: Particle A (P_A) – P25 TiO_2 (Evonik Degussa Corporation) and Particle B (P_B) – VHP-D TiO_2 (Altair). The particle sizes for P_A and P_B were approximately 20 nm and 35 nm in diameter, respectively, according to the manufacturers' data sheets. There is no surface treatment for P_A, while P_B contains an organic surface treatment. Polymeric matrices used were a solvent-borne acrylic urethane (AU) (Joncryl BASF and Demodur, Bayer) and water-borne butyl-acrylic styrene latex (UCAR 481, Dow Chemical).

Sample Preparation

The nanoparticles were incorporated into AU or latex coatings at a concentration of 5 % (particle volume concentration). The preparation of AU films has been described in previous studies [4]. The thickness of dry films is about 110 μm. The latex films were prepared according to latex E2216-C formulation, and the procedures can be found elsewhere [5]. The dry film thickness is about 80 μm.

Morphology and Nanoparticle Dispersion Characterization

Surface and subsurface morphology and nanoparticle dispersion in AU and latex coatings were characterized using a Zeiss model LSM510 reflection laser scanning confocal microscope (LSCM) with a laser wavelength of 543 nm. A detailed description of LSCM analysis for these samples can be found elsewhere [6]. LSCM images are presented in 3 forms: two dimensional (2D) projections in x-y plane, a side-view projection in the x-z plane, and a single depth-profile image at a one z-depth (depth-profile image). The 2D LSCM image (512 pixel × 512 pixel) is formed by summing the stack of images over the z direction (normal to coating surface). Side view image is formed by summing the stacks of images over y direction (lateral to coating surface). Pixel intensity level represents the sum of back-scattered light. Darker areas represent regions scattering less light than lighter colored areas. Single depth-profile image shows the microstructure and nanoparticle cluster distribution in the subsurface layers. Depth-profile subsurface images presented here are around 3.0 μm below the polymer-air surface for coatings containing P_A or P_B.

Surface Mechanical Properties

Near-surface mechanical properties were measured using a NanoXP depth sensing instrument (Agilent Technologies) equipped with two different tip geometries. A Berkovich pyramidal tip and a spherical tip (tip radius of 10 μm with a semi-apical angle of 45°) were used to indent the coatings. All indentation experiments were indented to a depth of 3.0 μm using a strain rate of 0.05 s^{-1}. The tip-sample contact stiffness [7] was measured by superposing a harmonic oscillation of 5 nm at a frequency of 45 Hz over the loading portion of the indent. The reported modulus values were the average of 20 indentations over a depth range from 1000 nm to 2000 nm without a drift correction.

RESULTS AND DISCUSSION

AU System

The dispersion of TiO_2 nanoparticles on the surface and subsurface in the AU matrix was characterized using LSCM. Figure 1 shows the 2D LSCM projections, single-layer depth-profile (3.0 μm below polymer-air surface) and side view projection for AU coatings containing P_A and P_B. The high backscattered intensity regions represent TiO_2 clusters with a higher index of refraction on or near the surface, while dark regions are associated with the AU matrix. This backscattered intensity is proportional to the 6th power of the cluster size, thus larger cluster sizes result in a higher intensity. In the 2D projection, there are very few bright spots on the surface of

AU-5 % P_B system in contrast to the surface of AU-5 % P_A system. This implies that there is a thicker clear polymer-rich layer [8] near the polymer-air surface for the AU-5 % P_B system and that most of P_B particles are below the surface and buried deep into the polymer matrix. These observations are also confirmed by the side view images. Notably, P_A clusters are distributed uniformly and tightly packed near the surface. The total scanning depth in the z direction (z-depth) was around 6 μm for AU-5 % P_A system. In comparison, the P_B clusters are larger and loosely packed so that the laser can penetrate deeper into the coating and hence the z-depth is larger. The total z-depth was approximately 13 μm for the AU-5 % P_B system. The nanoparticle dispersion (particle shape and size) in the coatings can be also quantified using depth-profiling (these subsurface images in Figure 1). There are smaller bright spots uniformly distributed in the AU-5 % P_A system. In contrast, there are fewer but larger bright spots distributed randomly in the AU-5 % P_B system. These clusters are as large as 10 μm. Ongoing image analyses on the particle dispersion will be reported elsewhere.

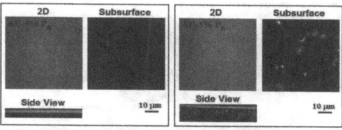

Figure 1. 2D LSCM projection, a single-layer subsurface (3 μm below polymer-air surface), and side view images for AU systems containing 5 % P_A and P_B. The scale bar is 10 μm.

The effect of nanoparticle dispersion on surface mechanical properties was investigated using DSI. Table 1 summarizes the indentation modulus (E) data for the AU systems. For both types of tips, E increases relative to pure AU with the addition of 5 % P_A or P_B; but a larger increase and standard deviation was observed with the addition of P_B. As mentioned previously, the AU-5 % P_A system has a smaller cluster size and better nanoparticle dispersion than the AU-5 % P_B system. A higher variability is expected in the indentation measurements on the AU-5 % P_B system due to the greater heterogeneity in cluster size and distribution. Reporting the average E from multiple indents provides a glimpse of the effect of the TiO_2 nanoparticles on the mechanical properties. However, it does not clearly reflect the heterogeneity of TiO_2 cluster distribution on the surface or subsurface of polymeric coatings.

Table 1. Surface modulus (E) of AU systems measured with DSI with two different tips. The $\pm$ values represent standard deviation from the averaged data of 20 indents.

	Pure AU (GPa)	AU - 5 % P_A (GPa)	AU - 5 % P_B (GPa)
Pyramidal	3.62 ± 0.04	3.73 ± 0.04	4.12 ± 0.41
Spherical	3.73 ± 0.04	3.98 ± 0.03	4.29 ± 0.36

To obtain the effect of local TiO_2 cluster distribution on surface modulus, E values were plotted as a function of TiO_2 cluster location and depth for each indent. Figure 2 shows residual indents for AU-5 % P_A system using a pyramidal tip and AU-5 % P_B system using both tips and their modulus-displacement (E-d) curves for four selected indents. There is little difference observed in the E-d curves in the AU-5 % P_A system. The residual indents appear similar with a symmetrical pyramidal shape (Figure 2a). Similar results (not reported here) were also found for AU-5 % P_A system using the spherical tip. However, for AU-5 % P_B systems, a few residual indents display an irregular shape (indents #1, #3 and #8 in Figure 2b) using a pyramidal tip. These extended or distorted corners of the pyramidal shape are near the location of the P_B clusters on or below the surface (Figure 2b, middle graph). For example, for a P_B cluster on the surface (indent #8), a higher E value was observed near the surface. While for indent #1 (or indent #3), the E values increase gradually with depth until the indenter encounters a P_B cluster inside the coating. The E-d curves for most of the indents appear similar to that of #13, indicating that most of the indents did not encounter the P_B clusters. To investigate the impact of tip geometries, the residual indents of AU-5 % P_B system using a spherical tip were also studied (Figure 2c). It is found that the changes in modulus as a function of displacement using a spherical tip are not as sensitive as a pyramidal tip.

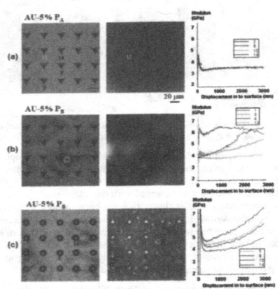

Figure 2. 2D projection (left column), a single-layer subsurface image (3 μm below polymer-air surface, middle column), and corresponding E-d curves (right column) for four selected indents as indicated in the 2D projection for AU systems containing 5 % P_A (a), P_B (b and c). (a) and (b) are DSI results using a pyramidal tip and (c) is DSI result using a spherical tip. The scale bar is 20 μm.

Latex System

Figure 3 shows the LSCM images of latex coatings containing P_A and P_B. Similar to the AU coatings, P_A clusters are more uniformly distributed on or near the surface than P_B clusters in the latex coatings. From the subsurface (3 μm below) and side view projection, it is confirmed that

P_A clusters are larger in the latex coating than that in the AU coating (Figures 1 and 3). Larger cluster sizes and a larger scanning z-depth (ca. 16 µm) were also observed in the latex-5 % P_B system. After examining many subsurface layers, it is concluded that the AU-5 % P_A system has a smaller cluster size and has the most uniform particle dispersion; the latex-5 % P_B system has the least uniform particle dispersion among the four systems. Note that better dispersion is defined as smaller cluster sizes that are more uniformly distributed throughout the measurement area.

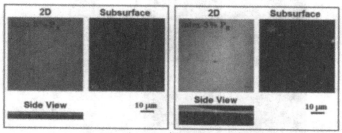

Figure 3. 2D projection and a single-layer subsurface (3 µm below surface) image, and side view image for latex systems containing 5 % P_A and P_B. The scale bar is 10 µm.

Table 2. Surface modulus (E) of latex systems measured with DSI with two different tips. The ± values represent standard deviation from the averaged data of 20 indents.

	Blank latex (GPa)	latex - 5 % P_A (GPa)	latex - 5 % P_B (GPa)
Pyramidal	0.88 ± 0.02	0.98 ± 0.05	1.17 ± 0.16
Spherical	0.95 ± 0.02	1.02 ± 0.03	1.30 ± 0.16

Table 2 summarizes indentation modulus (E) for the latex systems. Similar to the AU system, the average value and standard deviation of E for this data set increased significantly with the addition of P_B. The residual indents for latex-5 % P_A system using a pyramidal tip (Figure 4 a) and latex-5 % P_B system using both pyramidal and spherical tips (Figures 4 b and 4c) and their E-d curves for four selected indents are displayed in Figure 4. There is no significant difference in the E-d curves for all the indents in the latex-5 % P_A system (Figure 4 a). However, for latex-5 % P_B system (Figure 4 b), some residual indents show irregular shapes, and this result is similar to the findings for the AU-5 % P_B system using the pyramidal tip (Figure 2b). The E-d curves of latex-5 % P_B system using a pyramidal tip also show a good coincidence with particle cluster distribution, as shown in the indents #6, #15 and #19 in Figure 4b. However, the E-d curves of latex-5 % P_B system using a spherical tip appear to be similar, and this result implies that the DSI using the spherical tip is not as sensitive as the pyramidal tip for detecting the spatial distribution of nanoparticle clusters.

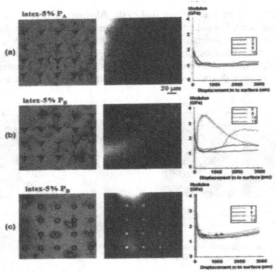

Figure 4. 2D projection (left column), a single-layer subsurface image (3 μm below polymer-air surface, middle column), and corresponding E-d curves (right column) for four selected indents as indicated in the 2D projection for latex systems containing 5 % P_A (a), P_B (b and c). (a) and (b) are DSI results using a pyramidal tip and (c) is DSI result using a spherical tip. The scale bar is 20 μm.

CONCLUSIONS

A method to investigate the effect of nanoTiO$_2$ cluster dispersion on the surface mechanical properties in two different nanoTiO$_2$-polymer systems was developed using a combination of laser scanning confocal microscopy and depth sensing indentation. Good coincidence was found for nanoparticle dispersion and surface modulus mapping using a pyramidal tip.

REFERENCES

1. ACS Symposium Series1008: *Nanotechnology Applications in Coatings*, Eds: R. H. Fernando, L. P. Sung (ACS/Oxford University Press. 2009).
2. S. Krieger, I. Cabrera, M. Ratering, T. Fichtner, R. Farwaha, JCT CoatingsTech., 26-30, (2008).
3. L. P. Sung, S. S. Watson, M. Baghai-Anaraki, D. L. Ho, Mat. Res. Soc. Symp. Proc., 740, 1541(2003).
4. D. L. Wang, S. S. Watson, L. P. Sung, I-H. Tseng, C.J. Bouis, R. Fernando, J. Coat. Technol. Res., (2009).
5. E. Flick, Water-Based Paint Formulation 4, Noyes Publication, Westwood, NJ, p34 (1994).
6. J. Faucheu, L.P. Sung, J. W. Martin, K. A. Wood, J. Coat. Technol. Res., 3, 29, (2006).
7. M. R. VanLandingham, J. of Res. of the Nat'l Inst. of Stds. and Tech., 108, 249 (2003).
8. C. Clerici, X. Gu, L. P. Sung, A. M. Forster, D. L. Ho, P. Stutzman, T. Nguyen and J. W. Martin, *Service Life Prediction for Polymeric Materials: Global Perspectives*, Eds: J. Martin, R. Ryntz, J. Chin, R. Dickie, Chapter 31, pp 475-492 (Springer Press, 2009)

Mater. Res. Soc. Symp. Proc. Vol. 1224 © 2010 Materials Research Society   1224-FF10-09

Atomistic simulations of the mechanical response of copper/polybutadiene joints under stress

F. O. Valega Mackenzie and B. J. Thijsse
Department of Materials Science and Engineering, Delft University of Technology
Mekelweg 2, 2628 CD Delft, The Netherlands

ABSTRACT

Metal/polymer system joints are widely encountered nowadays in microscopic structures such as displays and microchips. In several critical cases they undergo thermal and mechanical loading, with contact failure due to fracture as a possible consequence. Because of their variety in nature and composition metal/polymer joints have become major challenges for experimental, theoretical, and numerical studies. Here we report on results of molecular dynamics simulations carried out to study the mechanical response of a metal/polymer joint, in this case the Cu/polybutadiene model system. The behavior of Cu and the cross-linked polybutadiene are modeled, respectively, by the Embedded Atom Method (EAM) and the Universal Force Field (UFF). Loading is applied under compression. Different potentials are used to describe the interactions in the metal/polymer interface, which allows us to qualitatively analyze possible mechanisms of failure in these joints, below the metal melting point and above the polymer glass transition temperatures.

INTRODUCTION

Polymer/metal joints are widely encountered in electronic devices [1], ranging from microscopic parts such as displays and microchips to solar power cell devices and coatings found in macroscopic engineering applications. Because of this variety in composition and nature the interfaces of such joints constitute challenging systems for studies (both theoretical [2,3] and experimental [4]) of the behavior under tensile or compressive stresses, temperature variations and also in the determination of adhesion.

The purpose of this work is to take the first steps in setting up a consistent computational framework that permits to accurately model and study the mechanical behavior of polymer/metal interfaces at an atomistic level. As a first approach we use the Universal Force Field (UFF) [5] together with the Embedded Atom Method [6], for describing the atomic interactions in the polymer and the metal, respectively. For the time being the UFF is also used for the mixed interaction between the metal and polymer atoms. This last choice was made on the basis of the universality of UFF, which will allow us to simulate polymers with different compositions and at the same time contains the necessary parameters to model to a first approximation the interaction at the interface with the metal atoms.

COMPUTATIONAL MODELS

All simulations were carried out by the LAMMPS molecular dynamics code. Modifications to it were made in order to adapt the force field potential forms as encountered in the original UFF formulation for the angular, torsional and improper terms. The timestep chosen for each run was 0.1 fs because we are considering hydrogens explicitly. At this stage, no charges were included in the model. This can be justified to a certain extent by the absence of strongly electronegative atoms (such as oxygen).

In order to create the polymer/metal joint (Cu/polybutadiene) we first need to set up the polymer and metal phases. In the following we describe the way in which each part was constructed.

Polymer

The strategy adopted to build the cross-linked polybutadiene consisted of a variation of a procedure used in previous works [7] in which the polymer is built up bond-by-bond until it fills the required volume with the required density. In our case several polymer chains were grown from random starting locations inside the volume, and we added to this the building rule that if two carbon atoms belonging to different chains are within a cut off distance of 1.8 Å, a cross-linking bond between them is created. After the desired density of 0.8 g/cm^3 [8] was achieved, we proceeded to relax the structure in a three-step procedure. First we minimize the structural energy using the steepest descent method. Then we heat up the system through molecular dynamics runs from 0 K to a very high temperature (even above the melting point of the polymer), in this case 900 K. At this point we recall that the polymer remains topologically intact and does not melt in the proper way, because of the restrictions in the force field, which do not allow bonds to be broken. As a final step the system is again cooled down to a temperature of 300 K, resulting in a 1 ns duration of the total relaxation procedure.

Studies to study the dependency of the mechanical properties of polybutadiene constructed in this way on the number of cross-links are underway and will be reported in a forthcoming publication. It is important to note that once the polymer is cross-linked, no new bonds are created/destroyed. This ensures that the polymer will maintain its integrity irrespective of the pressure and temperature.

Metal

For the copper crystal we created a periodic fcc structure with a lattice constant of 3.615 Å, using the EAM potential as described in [9]. To relax the system an initial gaussian distribution of velocities was given and then a run for 100 ps in the NVT ensemble was performed. The Cu surface that was freely relaxed before interacting with the polymer was the (001) crystal face.

The interface

After relaxation of the metal and the polymer, the interface was created in the following way. The two systems, as shown in figure 1a, were brought together from a safe distance of 20 Å (twice the cut off distance for the non-bonded interaction, 10 Å in this case), and the polymer was pressed towards the metal over a repeated sequence of 2 steps, the movement of a wall located on the far side of the polymer, at a rate of 0.1 Å/ps, during 10 ps, followed by a relaxation of the system for another 10 ps with the wall kept immobile. This process was repeated for a total of 300 ps.

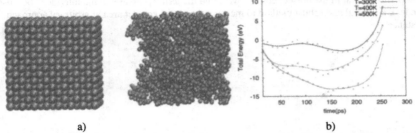

a)	b)

Figure 1. Initial set up before creating the interface. a) fcc copper crystal and cross-linked polybutadiene before the creation of the interface. The box sizes of the metal and polymer are both 30 x 30 x 30 Å. Copper atoms are in orange, carbon atoms in green and hydrogen atoms in gray. The interface was created later by bringing the two systems together by means of a wall pushing from the right. b) Time dependence of the total energy of the system when the surfaces are brought together at different temperatures.

This process was simulated at constant temperature and the significance of the wall is merely providing a mechanism to move the surfaces together. Also, the last 3 planes of the Cu crystal, on the side opposite to the interface, were held fixed in space.

RESULTS

Ensemble averages were taken each 5 ps, which is a sufficiently high frequency considering the procedure of approach that we have adopted. Figure 1b summarizes the total energy data for three temperatures. As expected we observe that (disregarding the wall oscillations) when both surfaces start to interact, the total energy of the system decreases. This continues until the energy reaches a minimum and starts to increase, showing no signs of any other minima. This shape of the total energy curve is basically that of a simple potential well, which is consistent with the fact that all non-bonded atomic interactions have a Lennard-Jones nature.

To validate the consequences of the interface formation we have plotted the internal pressure in the metal in figure 2, and we observe that just after the energy minimum is reached the pressure in the metal stops being negligible and begins to rise, confirming that the wall is no longer just bringing the two surfaces together but compressing them. It is also interesting to notice that for different temperatures the time it takes the system to create the interface varies, as well as the net gain in total energy of the system. For the higher temperatures studied (400K and 500K) there seems to be no significant difference in the time to reach equilibrium, although the strength of the attractive interaction increases with increasing temperature.

For the formation of the interface we preferred to compare the total energy, rather than the distance, as a function of time, because of the imprecise definition of the interaction distance between a crystalline and an amorphous surface.

Compression under load

The effect of applying a load was simulated by letting a net force act on all atoms in the polymer, perpendicular to the interface in the direction of approach. Forces were applied in the range 10-

100 pN. Note that in the present context we avoid the term "pressure on the interface" because we lack a general and robust method to measure the actual contact area at arbitrary approach distance.

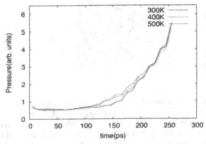

Figure 2. Internal pressure in the metal during the approach of the metal and the polymer at 3 different temperatures.

The intention of this compressive loading was to explore the behavior of the copper/polybutadiene joint as a whole after exposure to extreme mechanical conditions. With the current scheme we found that the integrity of the joint is maintained and the variation of the volume is independent of the temperature, as can be seen in figure 3a. Further calculations show that the metal is responsible for less than 10% of the contraction for the maximum load (100 pN), and less than 1% for 10 pN. So, as expected, the major part of the volume change is due to the deformation in the polymer.

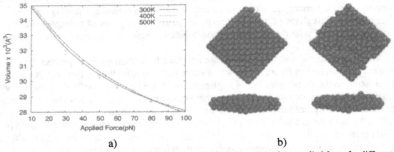

a)   b)

Figure 3. a) Variation of the volume of the Cu/polybutadiene joint as a function of the applied force for different temperatures. b) Left: metal surface at 10 pN load. Right: the same surface but at 100 pN load. Both are at the same temperature, 300 K.

Nevertheless, we find an interesting change in the roughening of the metal surface when the applied force exceeds a certain value (70 pN for 300 K), see figure 3b. This morphological variation was found after 1ns simulation, only in the first 2 atomic layers of the metal surface. The metal atoms in surface remained attached to the bulk, meaning that no diffusion of the metal atoms into the polymer was observed.

DISCUSSION AND CONCLUSIONS

An interface between copper and cross-linked polybutadiene was created and studied under loading by molecular dynamics. Although the Cu/polybutadiene mixed interaction was modeled in a fairly simple way, the bonding of the materials was energetically realistic (a couple of eV). The bonding was found to increase with increasing temperature. This might be caused by several factors. One possibility is that a higher temperature allows the atoms to relax faster and thereby increase their bonding more effectively. This explanation seems to be supported by the steeper slope in each case before the minimum energy is reached. A second possibility is that with a higher temperature the effective interface area and volume increase (due to thermal motion), therefore increasing the net interaction.

When the system is subjected to severe compressive conditions, the joint maintains its integrity after a load of 100 pN. Assuming that this force is homogeneously distributed over the interface we would end up with a pressure of around 160 MPa. With increasingly larger values for pressure and temperature one may expect non-linearity in the response [10], and very likely the formation of an interface with a mixed phase [4] between the metal and the polymer, as suggested by the changes in morphology obtained earlier. We plan to study such phenomena in the near future by the use of reactive potentials [11], which will help us to model metal/polymer systems in a more advanced way.

Finally, it is worth mentioning that there exist recent efforts [12] to implement methods to determine the surface area of the interaction in order to accurately calculate the pressure over the interface.

REFERENCES

1. W. D. van Driel, R. B. R. van Silfhout and G. Q. Zhang. *IEEE Trans. Device Mater. Reliability* 9, 53 (2009).
2. M. Deng, V. B. C. Tan and T. E. Tay. *Polymer 45*, 6399 (2004).
3. J. C. Suarez, S. Miguel, P. Pinilla and F. Lopez. *J. Adhes. Sci. Technol.* 22, 1387 (2008).
4. P. S. Ho, R. Haight, R. C. White and B. D. Silverman. *J. Physique – Colloque 5 Supp. 10*, 49 (1988).
5. A. K. Rappé, C. J. Casewit, K. S. Colwell, W. A. Goddard III and W. M. Skiff. *J. Am. Chem. Soc.* 114, 10024 (1992).
6. M. S. Daw and M. I. Baskes. *Phys. Rev. B* 29, 6443 (1984).
7. Y. Li and W. L. Mattice. *Macromolecules* 25, 4942 (1992).
8. R. A. Pethrick and R. W. Richards. *Static and Dynamic Properties of the Polymeric Solid State*. Dordrecht, Holland, 1982.
9. S. M. Foiles, M. I. Baskes and M. S. Daw. *Physical Review B* 33, 7983 (1986).
10. J. B. Hooper, D. Bedrov, G. D. Smith, B. Hanson, O. Borodin, D. M. Dattelbaum and E. M Kober. *J. Chem. Phys.* 130, 144904 (2009).
11. C. F. Sanz-Navarro, P. O. Åstrand, D. Chen, M. Rønning, A. C. T. van Duin, J. E. Mueller and Goddard III W. A. *J. Phys. Chem. A* 112, 12663 (2008).
12. Y. Mo, K. T. Turner and I. Szlufarska. *Nature* 457, 1116 (2009)

Simulations & Modeling

Mater. Res. Soc. Symp. Proc. Vol. 1224 © 2010 Materials Research Society   1224-GG05-15

MD Simulation of Dislocation Dynamics in Copper Nanoparticles

Yoshiaki Kogure, Toshio Kosugi, Tadatoshi Nozaki and Masao Doyama
Teikyo University of Science and Technology, 2525 Yatsusawa, Uenohara, Yamanashi 409-0193, Japan

ABSTRACT

Atomistic configuration and motion of dislocation have been simulated by means of molecular dynamics method. The embedded atom method potential for copper is adopted in the simulation. Model crystal is a rectangular solid containing about 140,000 atoms. An edge dislocation is introduced along [112] direction near the center of model crystal, and the system is relaxed. After the dislocation configuration is stabilized, a shear stress is applied and released. Wavy motion of dislocation is developed on the Peierls valleys when the free boundary condition is adopted. Motion of pinned dislocation is also simulated.

INTRODUCTION

The mechanical relaxation due to dislocation has been observed in copper, aluminum and other metals by means of internal friction measurements. Broad relaxation peaks observed at temperatures between 100 K and 200 K in the deformed samples are called Bordoni peak. If these relaxation peaks are analyzed based on the double kink formation mechanism, a larger value of Peierls stress, $10^{-3}\mu$, is derived, where μ is the shear modulus [1]. One of the present authors Kosugi discovered a new relaxation peak at 11 K in zone-refined aluminum samples, and derived Peierls stress was in the order of $10^{-5}\mu$, which is reasonable size as expected from the plastic deformation experiments [2]. On the other hand, we have performed a MD simulation of dislocation motion in 2-d model. In the simulation, the total kinetic energy is found to make a bump in the time variation, when the dislocation surmount a Peierls potential hill. Estimated Peierls stress is in the order of $10^{-5}\mu$ from the size of the hump [3]. The purpose of the present study is to develop a 3-d model for the simulation of dislocation in copper, and to investigate the structure and the motion of dislocations. A newly developed embedded atom method potential [4] is used in the present simulation. The potential function has successfully been applied on the simulation of the dynamics of crystal defects and nanoparticles [5-8]. Many studies of dislocations by MD simulation have been performed under the periodic boundary condition. To determine the Peierls stress a dislocation dipole of a straight line is introduced in a periodic super cell in the simulations to compensate the long range strain [9]. The fixed or the free boundary condition is adopted in the present simulation to realize the flexible dislocation appeared in the internal friction experiments.

METHOD OF SIMULATION

Molecular dynamics simulation has been performed by using an EAM potentials. The potential energy for the i-th atom is expressed as

$$E_i = F(\rho_i) + \sum_j \phi(r_{ij})/2, \tag{1}$$

where $F(\rho_i)$ is the embedding energy for the i-th atom and ρ_i is the electron density function, which is a sum of electron density of neighbor atoms labeled by j. These are expressed as

$$F(\rho_i) = D\rho_i \ln \rho_i, \tag{2}$$

$$\rho_i = \sum_j f(r_{ij}), \tag{3}$$

and $\phi(r_{ij})$ is the two body interaction between atoms i and j. The functional form of $\phi(r)$ and $f(r)$ are

$$\phi(r) = A(r_{c1} - r)^2 \exp(-c_1 r), \tag{4}$$

$$f(r) = B(r_{c2} - r)^2 \exp(-c_2 r), \tag{5}$$

where $r_{c1} = 1.6r_0$ and $r_{c2} = 1.9r_0$ are the truncation distance of the potential, and r_0 is the nearest neighbor distance. The potential parameters, A, B, D, c_1, c_2, for Cu are determined to reproduce the material properties such as the cohesive energy, the elastic constants, the vacancy formation energy and the stacking fault energy [4].

The time interval Δt for the simulation is chosen to be 1×10^{-15} s, which is less than 1/100 of the period of maximum atomic vibration frequency. The model crystal is covered by $\{110\}$, $\{111\}$, $\{112\}$ faces, and consisted of 143,000 atoms. Sizes are about $12 \times 10 \times 10$ nm in x, y, z directions, respectively. Fixed and free boundary conditions are adopted, namely, atoms on the surfaces perpendicular to the dislocation line are fixed or released free. The dislocation is pinned down at the surfaces in the former. On the contrary the dislocation can move freely in the latter. Atoms on the surfaces parallel to the dislocation line are fixed in both cases during simulation. Initially a couple of extra $\{110\}$ atomic planes are inserted in the lower middle of the crystal to introduce an edge dislocation because the fcc crystal is constituted by two independent planes in [110] direction, and the crystal is relaxed. The introduced dislocation is split to two partial dislocations as expected. A convenient way to express the position of dislocation is to highlight the atoms, which have higher potential energy than a certain criterion.

RESULTS AND DISCUSSION

Configuration after initial relaxation

Initially an edge dislocation is introduced. Configuration of atoms is shown in figure 1. (a), where atoms are displaced by the strain field given by the elasticity theory. The potential energy of each atom is calculated by equation (1), and the atoms with potential energy larger than -3.2 eV are highlighted by solid circles. Then the molecular dynamics simulation was performed and the crystal was relaxed. Result is shown in figure 1 (b). Introduced dislocation is seen to be splitted to partial dislocations. Vertical view of atomic planes is shown on right hand side of the figure. A stacking fault structure is seen between two partial dislocations. Width of stacking fault is known to be determined by the balance of the stacking fault energy and the repulsive force between two partial dislocations but the effect of fixed boundary may reduce the width of stacking fault.

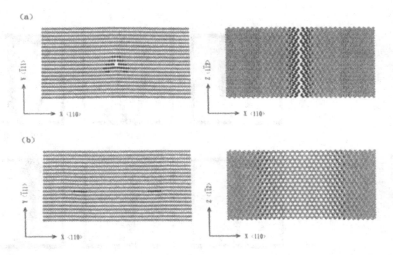

Figure 1. Configuration of atoms around an edge dislocation (a) before and (b) after relaxation under a fixed boundary condition. Solid circles express atoms with the potential energy, $E > -3.2$ eV. Only the central part of the crystal is shown.

Motion of split dislocations

A shear strain $\varepsilon_{xy} = 0.05$ to move the dislocation in $-x$ direction was applied as a step function of time and system was relaxed under the fixed boundary condition. Then the motion of dislocation was monitored. The simulation was performed at the temperature, 0 K, but it was slightly increased up to 6 K with the molecular dynamics (MD) time steps. Snap shots of the dislocation configuration at typical MD time steps are shown in figure 2, where only atoms with higher potential energy, $E > -3.5$ eV, are shown, and the atoms with the potential energy higher than -3.485 eV are marked by solid circles. These open and solid circles characterize the internal structure of the dislocations. In figure 2, the dislocations are pinned at top and bottom surfaces by adopting the fixed boundary condition. Initially one of the partial dislocation moves toward left direction ($S = 2000$), then another one follows. Both dislocations form bow shape as a balance of external stress and line tension of dislocations.

To investigate the dislocation dynamics in detail the atomistic configuration at typical time steps are magnified and shown in figure 3, where atoms with larger potential energy ($E > -3.5$ eV) are shown. We consider these atoms constitute a dislocation. In the atomistic model dislocation lines is spread in several atomic lines and they move simultaneously. Figure 4 shows the change of potential energy of dislocation atoms ΔP during the dislocation motion. ΔP shows clear sharp minima at time steps marked by arrows A--D. At the time right-side end

127

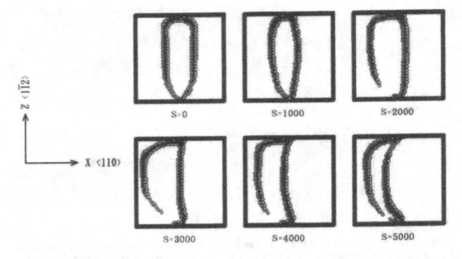

Figure 2. Motion of split edge dislocation under a fixed boundary condition. S represents MD time steps. Open circles express atoms with the potential energy, -3.5 eV $< E < -3.485$ eV, and solid circles express atoms $E \geq -3.485$ eV.

(a)   (b)

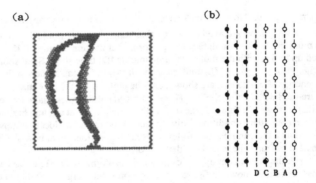

Figure 3. (a) Atoms constituting partial dislocations at MD step 3500 are shown by open circles and position of the atoms at 4000 step are overwritten by solid circles. (b) A part surrounded by a rectangle in figure 3 (a) is magnified and is shown in figure 3(b). Initially, the right-hand side of the dislocation was at the line O and moved by one line. The change of potential energy during the motion is shown in figure 4.

128

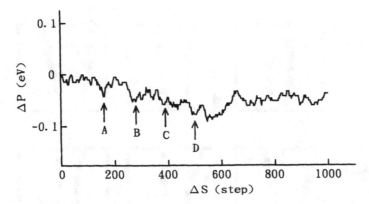

Figure 4. Change of total potential energy of split edge dislocation under a fixed boundary condition. ΔS represents additional step to 3500 MD time step. ΔP shows minima at the points A, B, C, D, where the dislocation is in stable positions (see figure 3).

stays the position shown by dashed lines, A—D, in figure 3, where dislocation seems to be stable. Then the broad peaks between arrows in figure 4 may be due to the Peierls potential. The Peierls potential for a dislocation stays along the z -direction is periodic in the x -direction and shows minima at the positions of dashed lines in figure 3(b) (the $x-$ and the $z-$ directions are the same as figure 2) . Short and quick fluctuations of the order of 0.02 eV are seen to be superposed on the broad peak by the Peierls potential in figure 4. These fluctuations are anticipated to be energy change due to kink motion. As seen in figure 3 (b), there is the point where one atom protrudes in the neighboring valley in spots not in alignment, and these may correspond to the atomistic structure of a kink. Conventionally, the energy for the kink motion has been discussed on the basis of the Peierls-Nabarro model. A detailed comparison with the atomistic simulation and the calculation by the Peierls-Nabaro model should be made in future.

Results for the simulation under the free boundary condition are shown in figure 5. In this case the dislocation is not fixed at the top and the bottom surfaces, and is free to move in x- direction. The shapes of the dislocations are very wavy and many kinks seem to be present on the Peierls hills, which run in vertical direction on a slip plane. The configurations shown in figure 5 are not symmetrical in z- direction. It is noted that six distinct types of atomic planes are stacked upon one another in a repeating sequence as ABCDEFABCDEF...... in [112] direction, a configuration that is not symmetric in the top and bottom directions. Initially, the partial dislocation shown on the left side moved towards the left-hand side on applying the stress, afterwards it was pushed back by the mirror force due to the fixed boundary. These motions are repeated. A period of the dislocation oscillation is about 100 times of the thermal lattice vibration. As the dislocations are widely spread on a [111] plane, they are always confined in an atomic plane, namely, climbing is not observed in the simulation.

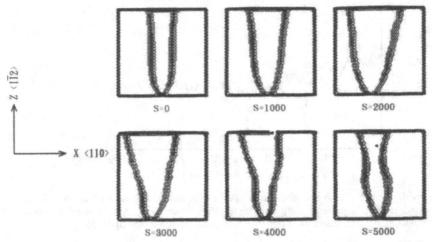

$Z \langle \bar{1}\bar{1}2 \rangle$

$X \langle 110 \rangle$

S=0   S=1000   S=2000

S=3000   S=4000   S=5000

Figure 5. Motion of split edge dislocation under a free boundary condition. S represents MD time steps. Open circles express atoms with the potential energy , -3.5 eV $< E < -3.485$ eV , and solid circles express atoms $E \geq -3.485$ eV .

CONCLUSIONS

EAM potential has successfully been applied on the molecular dynamics simulation of moving dislocations in a copper crystal. A dislocation was splitted into partials and a stacking fault appeared between partials. Motion of splitted partial dislocations are displayed by highlighting the atoms with larger potential energy. Wavy motion of partial dislocations on a $\{111\}$ plane was observed under the fixed and free boundary condition.

REFERENCES

1. D. H. Niblet, in *Physical Acoustics*, **Vol. IIIA**, edited by W. P. Mason (Academic Press, New York, 1966) p. 77.
2. T. Kosugi and T. Kino, *J. Phys. Soc. Jpn.* **58**, 4269 (1989)
3. Y. Kogure and T. Kosugi, *J. Phys. IV* (France) **Vol. 6**, pp. C8-195-198 (1996)
4. M. Doyama and Y. Kogure, *Radiat. Effects Defects Solids*, **142**, 107-114 (1997)
5. Y. Kogure, *J. Alloys Comp.*, **355**, 188-195 (220)
6. Y. Kogure, T. Kosugi and M. Doyama, *Mater. Sci. Eng.* **A370**, 100-104 (2004)
7. Y. Kogure, T. Kosugi, M. Doyama and H. Kaburaki, *Mater. Sci. Eng.* **A442**, 71-74 (2006)
8. Y. Kogure, T. Kosugi, T. Nozaki and M. Doyama, *Mater. Sci. Eng.* **A521-522**, 30-33 (2009)
9. V. V. Bulatov and W. Cai, Computer simulations of dislocations (Oxford University Press, Oxford, 2006) p. 96.

Mater. Res. Soc. Symp. Proc. Vol. 1224 © 2010 Materials Research Society   1224-GG07-03

Size Effect in the Shear-Coupled Migration of Grain Boundaries Pinned by Triple Junctions

Javier Gil Sevillano , Aitor Luque , Javier Aldazabal , José M. Martinez-Esnaola
CEIT and TECNUN, University of Navarra, M. de Lardizabal 15, 20018 San Sebastian, Spain.

ABSTRACT

This paper presents molecular dynamics simulations of shear-coupled migration of tilt boundaries pinned by triple junctions in a simple model structure of columnar grains of different sizes. Simulations are for copper at 300 K. The phenomenon is of interest as a possible explanation of the Hall-Petch relationship breakdown in nano-grained polycrystals deformed at high or moderate strain rate and low temperature.

INTRODUCTION

In a stressed polycrystal, tilt boundaries under resolved shear stress experience a virtual pressure that promotes their shear strain-coupled migration (SCM), like similar tilt boundaries in bicrystals do [1-21]. However, in polycrystals the grain boundaries are constrained in their motion by their limiting triple junctions. The pinning effect of the junctions induces the bulging of migrating boundaries that cannot take place without additional accommodation by elastic distortion (for early bulging) or in general by elastoplastic accommodation, with dislocation emission or grain boundary sliding (GBS) simultaneous to SCM [22, 23]. Finite migration of pinned boundaries by bulging requires an increment of the applied resolved shear stress on the boundary above the shear stress needed for the migration of a free-to-move flat boundary. Such increase depends on the size of the boundary. This paper presents molecular dynamics simulations of shear-coupled migration of tilt boundaries pinned by triple junctions in a simple model structure of columnar grains of different sizes. Simulations are for copper at 300 K. The phenomenon is of interest as a possible explanation of the Hall-Petch relationship breakdown in nano-grained polycrystals deformed at high or moderate strain rate and low-temperature [23], Fig. 1. For copper the HP breakdown and the threshold stress for SCM both occur for equivalent tensile stresses of the order of 0.9 GPa. For many tilt boundaries SCM occurs in preference to GBS at low temperature and high or moderate shear strain rate, as attested by experimental or numerical simulations [1-23].

SIMULATION TECHNIQUE

Molecular dynamics (MD) simulations of shear deformation of copper bicrystals and multicrystals of different sizes have been performed using the embedded atom method (MD-EAM) as previously described [22, 23]. The potential used was the Mishin potential [24]. The bicrystals were of uniform thickness $Y = 2.17$ nm in the y // [001] direction and of rectangular xz section varying from 2.12 x 4.01 nm^2 to 14.90 x 29.22 nm^2 ($X = Z/2$). The bicrystals were constructed with a symmetrical $\Sigma 17(530)$ tilt boundary lying on the xy midplane. The characteristics of the boundary are given in Table 1. The shear coupling factor β represents the shear displacement in x direction contributed by the migration of the boundary per unit distance in z direction under a positive σ_{xz} shear stress. Thus, a negative value of β means that under a positive shear stress $\sigma_{xz} > 0$, the boundary migrates in the negative sense of the z direction and the volume swept by the boundary in the adjacent grains undergoes a shear strain of $(-\beta)$.

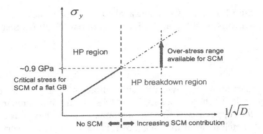

Figure 1. Sketch of the possible Hall-Petch law breakdown induced by activation of SCM. When the yield stress predicted by the Hall-Petch law reaches the critical stress for shear-coupled migration of free flat tilt boundaries (above some critical grain size D_{crit}) SCM starts.

The multicrystals (Fig. 2), also of thickness $Y = 2.17$ nm in the [001] direction, consisted of columnar grains with boundaries parallel to the thickness direction, with two central grains of orientations similar to those of the bicrystalline samples, plus two adjoining grains separated from the central pair of grains by grain boundaries inclined at $\pm 45°$ and emanating symmetrically from the xz (horizontal in the scheme) boundary. Their dimension in horizontal direction x ($X = 3Z/2$) ranged from 6.2 nm to 44.5 nm. The symmetrical $\Sigma 17(530)$ tilt boundary separating the upper and lower grains, of length $Z/2$, is now confined between two triple GB junctions. The common orientation of the two adjoining grains shares also the y // [001] direction with the central grains. Its boundaries are thus [001] tilt boundaries too. Their orientation is tilted $30.95°$ from that of the lower central grain. The $45°$ inclination of these boundaries has been chosen in order to minimize the resolved shear stress applied on them under a σ_{xz} shear stress applied to the samples. This will indeed only work for small strains (small rotations of the inclined boundaries).

Table 1. Characteristics of the xy (horizontal) boundary of the MD simulations.

Tilt boundary	Tilt axis	Tilt angle θ	Misorientation	Shear coupling factor β
$\Sigma 17(530)$	[001]	$61.9°$	$28.1°$	-0.5

Prior to virtual testing the samples were constructed at 0 K. Periodic boundary conditions in the three directions were assumed and they were relaxed during 5 ps at 0 K, temperature was linearly increased during 7 ps to 300 K and the samples were then relaxed during 12.5 ps at 300 K under no constraints for the GB to acquire its metastable configuration. After the final relaxation, two rigid zones 0.55 nm thick were established in the upper and lower layers of the samples. During the virtual shear test, the lower rigid zone remained fixed, the upper part being horizontally displaced at constant rate equivalent to a mean shear strain rate of the sample of 10^8

s^{-1}. Periodic boundary conditions were set along the x and y axes during shear deformation simulations. Atomistic configurations are viewed using AtomEye software [25].

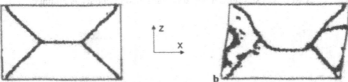

Figure 2. a) Multicrystalline sample ($X = 31.66$ nm) with columnar grains separated by tilt boundaries after relaxation at 300 K. **b)** The same sample after a mean shear strain $\gamma_{xy} = 0.11$ at 10^8 s^{-1}. FCC atoms are made invisible.

RESULTS

The macroscopic shear stress – shear strain behaviour of both types of samples is depicted in Fig. 3. The bicrystals display sawtooth curves characteristic of stick-slip behaviour, particularly for small size specimens or lower strain rates. Each serration represents the normal displacement H of the boundary from its equilibrium position to the nearest equivalent GB plane position, that implies a coupled displacement of the upper grain parallel to the GB plane $S = \beta H$ and provokes a stress drop. The drop is followed by a stress increase, a new normal displacement of the GB when the critical stress is reached, etc., and the process repeats. The transition between two stable positions requires small mechanically-driven correlated atomic displacements in the boundary without recourse to diffusion [15-17, 24]. The peak or critical shear stress for the $\Sigma17(530)$ boundary at 300 K and in the range of displacement rates used here (from 0.4 ms^{-1} to 4 ms^{-1}) varies from 0.2 GPa to 0.6 GPa. The columnar multicrystals show stress-strain curves that peak up to much higher stresses, of the order of 3 GPa, and then suffer a big stress drop followed by stick-slip with sub-peaks between 1 GPa and 2 GPa. In the first part of the curves, before reaching the maximum stress, repetitive small stress drops are evident and the curve, that reaches the peak at a shear strain $\gamma \approx 0.1$, departs from linearity.

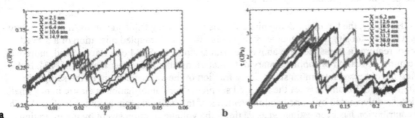

Figure 3. Shear stress – shear strain curves of respectively **(a)** bicrystals and **(b)** multicrystals at 10^8 s^{-1} shear strain rate. Notice scale differences.

In the case of the bicrystals, free coupled jerky SCM occurs under a critical stress dependent on the displacement rate. The area swept by the migrating boundary is uniformly deformed by shear according to its shear coupling factor, all in agreement with published results

[15-17, 20]. In the case of the multicrystals, an analysis of the images of the deforming samples shows that, before the maximum stress in the stress-strain curves is attained, shear coupled migration of the horizontal boundary has taken place. The boundary starts to migrate once the critical stress for SCM of the similar flat and unconstrained boundary is attained.

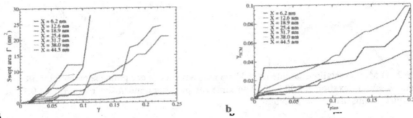

a
b

Figure 4. a) Area swept by SCM of the central (originally horizontal) tilt GB as a function of the total macroscopic shear strain of the samples. **b)** Estimated macroscopic shear strain contributed by SCM of the central GB, γ_{SCM}, to the macroscopic plastic shear strain, $\gamma_{plast} = (\gamma - \tau/G)$ with G the shear modulus. Macroscopic shear strain rate $\dot{\gamma} = 10^8$ s^{-1}.

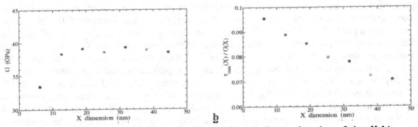

a
b

Figure 5. a) Macroscopic shear modulus G of the multicrystals as a function of size X. **b)** Maximum shear stress normalized by the size-dependent shear modulus G as a function of size X. Macroscopic shear strain rate $\dot{\gamma} = 10^8$ s^{-1}.

However, the boundary being pinned by its two limiting triple junctions, it needs to bow-out as it moves, loosing its flat geometry (e.g., Fig. 2b). Shear coupled migration of a curved boundary requires a stress increase above the critical stress for SCM of a flat one and, moreover, also requires activation of supplementary deformation mechanisms other than SCM [23]. Before reaching the observed maximum stress, a big fraction of the anelastic strain of the sample is contributed by SCM and most of the required supplementary deformation is elastic in our small samples, complemented by some local movements of the inclined grain boundaries (Fig. 4). The SCM contribution has been estimated as $|\beta|$ times the volume fraction swept by the migrating boundary. Finally, at the peak stress, nucleation of partial dislocations loops (most often from one of the triple boundaries) and their long range propagation takes place, accompanied by a big stress drop, Fig. 3b. Beyond that point, deformation occurs by intermittent partial dislocation emission and propagation with simultaneous SCM episodes (always in a jerky manner). The first peak stress is not reached again as deformation proceeds. Dislocation nucleation from the distorted grain boundaries seems to be easier now.

There is a size effect on the shear modulus, G, calculated from the first 0.002 deformation segment of the stress-strain curves. For sizes $X < 20$ nm the modulus weakens, Fig. 5a. The size effect is very strong on the peak stress, Fig. 5b, and it is of the smaller-the harder type. However, the size dependence of the maximum shear stress is not of the Hall-Petch type, it varies approximately as $X^{1/4}$ if a power law is fitted to the data. An exponential relationship starting from the critical stress for bicrystals fits better the data and leads to a saturation value for very small multicrystalline sample sizes, Fig. 6.

DISCUSSION AND CONCLUDING REMARKS

In Fig. 6 the maximum shear stress normalized by the size-dependent shear modulus has been plotted against $X^{1/2}$. The first data point corresponds to the bicrystals. In the same figure, the Hall-Petch relationship of coarse- to nano-grain copper at room-temperature [26] (valid down to about 20 nm, transformed to equivalent shear stress and normalized by the polycrystalline shear modulus), as well as the ideal shear stress for copper according to refs. [27, 28], 4.0 GPa (unrelaxed {111}<112> shear) have been plotted.

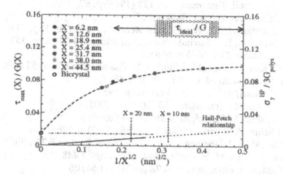

Figure 6. Maximum shear stress (normalized by the size-dependent shear modulus) calculated for multicrystal samples vs. inverse of square root of size, $X^{1/2}$. Macroscopic shear rate: 10^8 s^{-1}. The full line corresponds to the room-temperature Hall-Petch relationship of copper.

The presumed start of SCM in the polycrystals (the first data point in Fig. 6) occurs at a stress of the order reported for the Hall-Petch breakdown, for grain sizes of about 20 nm. Once the flow stress of a polycrystal has approached such value, deformation by SCM contributes to the plastic strain, in substitution of part of the dislocation mechanisms responsible for Hall-Petch strengthening. Moreover, as SCM represents grain growth, plastic yielding destabilizes the microstructure and some degree of Hall-Petch inversion is predicted. For the boundary studied, SCM occurs before grain boundary sliding is detected. Given the enormous gap between the SCM start stress in multicrystals and the maximum peak stress, it could be argued that the SCM contribution to strain rate and structural change will be irrelevant for Hall-Petch breakdown on account of the small available increments above the SCM start stress below the extrapolated

135

Hall-Petch line. However, it must be taken into account that the SCM results of Fig. 6 correspond to a 10^8 s^{-1} shear strain rate while the Hall-Petch results correspond to about 10^{-3}s^{-1}. SCM depends on temperature and strain rate according to a typical kinetic equation for jerky flow [17] with, for a $\Sigma 13(320)$ boundary, an activation energy of 1.7 eV. Assuming a similar activation energy for our $\Sigma 17(530)$ boundary, the strain rate contribution of SCM under the shear stresses determining the Hall-Petch breakdown could be relevant enough.

ACKNOWLEDGMENTS

Work supported by the Basque Government (ETORTEK inanoGUNE), Spanish Ministry of Science and Innovation and European Social Fund (Torres Quevedo Programme, A. Luque).

REFERENCES

1. J. Washburn, E.R. Parquer, *J. Metals-AIME Trans.* **4** (1952) p. 1076.
2. C.H. Li, E.H. Edwards, J. Washburn, E.R. Parker, *Acta Metall.* **1** (1953) p. 223.
3. D. McLean, *Nature* **172** (1953) p. 300.
4. S.G. Khayutin, *Phys. Met. Metallogr.* **37** (1974) p. 161.
5. M. Guillopé, J.P. Poirier, *Acta Metall.* **28** (1980) p. 163.
6. W.D. Means, M.W. Jessell, *Tectonophysics* **127** (1986) p. 67.
7. M.W. Jessell, *J. Struct. Geology* **8** (1986) p. 527.
8. M. Winning, G. Gottstein, L.S. Shvindlerman, *Acta Mater.* **49** (2001) p. 211.
9. M. Winning, G. Gottstein, L.S. Shvindlerman, *Acta Mater.* **50** (2002) p. 353.
10. M. Winning, *Acta Mater.* **51** (2003) p. 6465.
11. S. Zaefferer, J.C. Kuo, Z. Zhao, M. Winning, D. Raabe, *Acta Mater.* **51** (2003) p. 4719.
12. J.W. Cahn, J.E. Taylor, *Acta Mater.* **52** (2004) p. 4887.
13. F. Sansoz, J.F. Molinari, *Acta Mater.* **53** (2005) p. 1931.
14. M. Winning, A.D. Rollett, *Acta Mater.* **53** (2005) p. 2901.
15. J.W. Cahn, Y. Mishin, A. Suzuki, *Acta Mater.* **54** (2006) p. 4953.
16. J.W. Cahn, Y. Mishin, A. Suzuki, *Philos. Mag.* **86** (2006) p. 3965.
17. Y. Mishin, A. Suzuki, B.P. Uberuaga, A.F. Voter, *Phys. Review B* **75** (2007) p. 22401.
18. S. Badirujjaman, X.W. Li, M. Winning, *Mater. Sci. Eng. A* **448** (2007) p. 442.
19. V.A. Ivanov, Y. Mishin, *Phys. Review* B **78** (2008) p. 064106.
20. H. Zhang, D. Dhu, D.J. Srolovitz, *Philos. Mag.* **88** (2008) p. 243.
21. T. Gorkaya, D.A. Molodov, G. Gottstein, *Acta Mater.* **57** (2009) p. 5396.
22. A. Luque, J. Aldazabal, J.M. Martínez-Esnaola, J, Gil Sevillano, *Phys. Status Solidi C* **6** (2009) p. 2107.
23. A. Luque, J. Aldazabal, J.M. Martínez-Esnaola, J, Gil Sevillano, *Philos. Mag.* 2009 (in press).
24. Y. Mishin, M.J. Mehl, D.A. Papaconstantopoulos, A.F. Voter, J.D. Kress, *Phys. Rev. B* **63** (2001) p. 224106/1.
25. J. Li, *Model. Simul. Mater. Sci. Eng.* **11** (2003) p. 173.
26. N. Hansen, *Adv. Eng. Mater.* **7** (2005) p. 815.
27. D. Roundy, C. R. Krenn, M. L. Cohen, J. W. Morris Jr., *Phys. Rev. Letters* **82** (1999) p. 2713.
28. C. R. Krenn, D. Roundy, J. W. Morris Jr., M. L. Cohen, *Mater. Sci. Eng. A317* (2001) p. 44.

Mater. Res. Soc. Symp. Proc. Vol. 1224 © 2010 Materials Research Society 1224-GG05-03

Towards a Virtual Laboratory for Grain Boundaries and Dislocations

Sebastián ECHEVERRI RESTREPO[1] and Barend J. THIJSSE
Delft University of Technology, Department of Materials Science and Engineering, The Netherlands.

ABSTRACT

In order to perform a systematic study of the interaction between grain boundaries (GBs) and dislocations using molecular dynamics (MD), several tools need to be available. A combination of computational geometry and MD was used to build the foundations of what we call a virtual laboratory. First, an algorithm to generate GBs on face-centered cubic bicrystals was developed. Two crystals with different orientations are placed together. Then, by applying "microscopic" rigid body translations along the GB plane to one of the crystals and removing overlapping atoms, a set of initial configurations is sampled and a minimum energy configuration is found. Second, to classify the geometry of the GBs a local symmetry type (LST) describing the angular environment of each atom is calculated. It is found that for a given relaxed GB the number of atoms with different LSTs is not very large and that it is possible to find unique geometrical patterns in each GB. For instance, the LSTs of two GBs having the same "macroscopic" configuration but different "microscopic" degrees of freedom can be dissimilar: the configurations with higher GB energy tend to have a higher number of atoms with different LSTs. Third, edge dislocations are introduced into the bicrystals. We see that full edge dislocations split into Shockley partials. Finally, by loading the bicrystals with tensile stresses the edge dislocations are put into motion. Various examples of dislocation-GB interactions in Cu are presented.

INTRODUCTION

Within the framework of a multiscale project to model metallic interfaces and plasticity, the local characterization of the geometry of grain boundaries (GBs) and their discrete interaction with dislocations play a dominant role.

In the present paper we present our first steps towards the study of the local geometry of GBs and the interaction of a single dislocation with different GBs on face-centered cubic (fcc) materials. Additionally, the methodology used to generate bicrystals is shown.

A new way of locally characterizing GB geometry is introduced. A local symmetry type that quantifies the angular environment of each atom is calculated. It is found that, in general, for the minimum energy configuration of a given GB, repetitive patterns exist. These patterns tend to disappear for higher energy configurations.

Also, by applying a stress to a bicrystal containing a dislocation, the phenomena of absorption, nucleation and transmission of a single dislocation through a GB are investigated.

GENERATION OF GRAIN BOUNDARIES

GBs in cubic materials are usually described using five "macroscopic" degrees of freedom (DOFs). One of the methods used to account for these DOFs is the interface-plane

---

[1] s.echeverrirestrepo@tudelft.nl

scheme [1]. Here, the first four DOFs are defined by the Miller indices (MIs) of the two crystals. The fifth DOF consists of a rotation about the normal to the GB plane. Following this scheme, the DOFs of cubic crystals can be specified as $\{DOF's\} = \{(hkl)_1,(hkl)_2,\theta\}$.

Apart from this "macroscopic" description, three more "microscopic" DOFs are needed to fully define a GB [1]. These last DOFs correspond to small rigid body translations of one of the crystals and have a direct influence on the final energy and local geometry of the relaxed GB.

The procedure used to generate bicrystals is similar to that proposed in [2-4]. It consists of the following steps:

First, two periodic crystals with the desired MIs are placed together. Since having periodic boundary conditions along the GB plane is convenient for MD simulations, the periodicity of the two crystals has to be coincident with the MD simulation box. This restricts the number of different GBs that can be studied, but the MD box can be enlarged should the need arise. Note that periodicity in the direction perpendicular to the GB is no longer applied.

Second, a number of initial configurations are produced by applying several "micro-translations" (displacements of the order of 1/8 of the lattice parameter) to one of the crystals along the GB plane.

Third, since some of the atoms of the two crystals might be too close to each other near the GB plane [5, 6], a removal procedure is carried out. Atoms are discarded following four alternative criteria: 1) only atoms from the first crystal are removed, 2) only atoms from the second crystal are removed, 3) both atoms are removed or, 4) the atom to be removed is chosen randomly. Considering that the number of atoms on each configuration is usually different, this is an indirect way of varying the local density of atoms [7] and studying the effects of it. From each of these criteria one new configuration is obtained.

Finally, each configuration is relaxed using molecular dynamics (CAMELION [8]) without any constraints. The simulation starts with all the atoms having a kinetic energy corresponding to a temperature of half the melting point of the material. Then, the atoms are left free to interact with each other to find their most energetically favorable position while the temperature is diminished to 0 K. Constant pressure (0 Pa) is imposed and the simulation runs until the total energy stays constant.

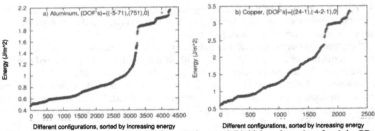

Figure 1. Energies after relaxation of different "microscopic" initial configurations for a) the GB $\{(\overline{5}71),(751),0\}(\Sigma25)$ on Al and b) the GB $\{(24\overline{1}),(\overline{42}\,\overline{1}),0\}(\Sigma7)$ on Cu.

It has been found that for GBs with the same "macroscopic" but different "microscopic" DOFs and/or atom removal criteria, the energy can be considerably different [2-4, 9]. The cases of the GBs $\{(\overline{5}7 1),(751),0\}$ ($\Sigma 25$) in Al and $\{(24\overline{1}),(\overline{42}\,\overline{1}),0\}$ ($\Sigma 7$) in Cu are presented in Figure 1. In both cases the potentials developed by Mishin et al. are used [10, 11].

LOCAL SYMMETRY TYPE

In order to characterize the local geometry of the GBs, a local symmetry type (LST) is assigned to each atom i. For this, twelve numbers Q describing its angular environment are calculated according to the following expression,

$$Q_i^{(\lambda)} = \frac{1}{Z_i}\sqrt{\sum_k^{Z_i}\sum_j^{Z_i}P^{(\lambda)}\left(\cos\left(\theta_{kij}\right)\right)}, \quad \lambda = 1,2,\cdots,12 \tag{1}$$

where $P^{(\lambda)}$ is the Legendre polynomial of degree λ, Z_i is the coordination number of the atom i and θ_{kij} is the angle formed by the triplet of atoms k, i and j, subtended at i. The $Q_i^{(\lambda)}$ values characterize bond orientations around atom i. They have been previously used to calculate an orientational order parameter to determine bond orientational order in liquids and gases [12] and to detect and distinguish the angular environment of atoms on MD simulations [13, 14].

Once these numbers Q are determined, the RMS difference $\left(\Delta LST_{i,j}\right)$ between each atom pair i, j is calculated:

$$\Delta LST_{i,j} = \sqrt{\frac{1}{12}\sum_{\lambda=1}^{12}\left(Q_i^{(\lambda)} - Q_j^{(\lambda)}\right)^2} \tag{2}$$

In this way, $\Delta LST_{i,j}$ is a single parameter that expresses the difference in bond orientations around atoms i and j. If its value is smaller than 0.09, the atoms i and j are classified as having the same LST. This critical bound was established by numerical experimentation.

By applying this approach on various bicrystals, it has been found that the number of atoms with dissimilar LSTs is generally not very large and that the minimum energy configuration of a given GB consists of atoms organized in repetitive geometrical patterns.

Two configurations of the GB $\{(031),(0\overline{3}1),0\}$ ($\Sigma 5$) with different energies are shown on Figure 2 [15]. In this figure, each color represents a different LST. The upper image corresponds to the minimum energy configuration found (0.832 J/m^2, 15168 atoms, height 90.87 Å, width 45.23 Å, depth 41.69 Å); it can be seen that the GB follows a repetitive pattern and that there are only 4 different LSTs. The lower image corresponds to a configuration with a higher energy (0.944 J/m^2, 15084 atoms, height 90.87 Å, width 45.38 Å, depth 41.81 Å) and it consists of atoms with 17 different LSTs. This illustrates a general tendency of the number of LSTs to increase for configurations with the same "macroscopic" DOFs but higher energy.

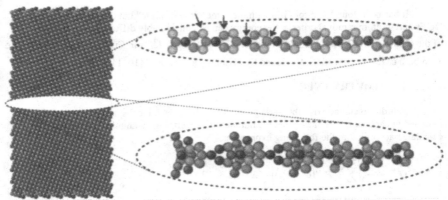

Figure 2. Two different local configurations of the GB $\{(031),(0\bar{3}1),0\}$ on copper. Atoms with different LSTs have different color. The upper GB has an energy of $(0.832\ J/m^2)$ and 4 different LSTs are present. The lower one has an energy of $(0.944\ J/m^2)$ and 17 different LSTs are present.

GRAIN BOUNDARIES AND DISLOCATIONS

In face centered cubic (FCC) materials like copper and aluminum, the most common edge dislocations are defined by a dislocation line $[\bar{1}12]$ and a Burgers vector $[\bar{1}\bar{1}0]$, [16]. To generate a full dislocation, two half planes (with a normal parallel to the direction $[110]$) are removed from a perfect crystal. Atoms are then shifted according to the displacement field predicted by the elastic isotropic theory of dislocations [16] and the whole system is relaxed at constant pressure (0 Pa) and temperature (0 K) using MD; note that since a linear defect is introduced, the only possible periodic direction is the one parallel to the dislocation line. At the end, as a consequence of the relaxation, the inserted dislocation splits into two Shockley partials. The application of this procedure to the minimum energy configuration of the GB $\{(24\bar{1}),(\overline{42}\,\bar{1}),0\}$ ($\Sigma7$) in copper is displayed in Figure 3. The coloring of the atoms is related to their local atomic environment: light blue represents fcc and yellow hcp. The system contains 120081 atoms (height 297.83 Å, width 184.89 Å, depth 34.68 Å) and the dislocation line defines the only periodic direction $[\bar{1}12]$, which is perpendicular to the image plane.

In order to force the dislocation to move towards the GB, the system is then compressed in the direction perpendicular to the GB plane (Figure 3). For this, the first three rows of atoms on the top and bottom free surfaces parallel to the GB plane are harmonically bound to anchor points prior to the start of the compression. Starting at $t = 0$, the anchor points on one side are pulled towards the other at a steady rate of 0.81 m/s. As the deformation increases, the dislocation approaches the GB, where it is fully absorbed (Figure 3a, b and c).

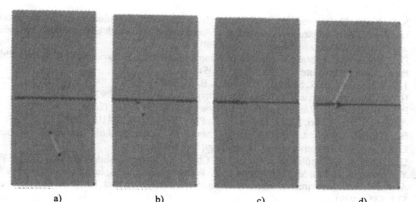

a)   b)   c)   d)

Figure 3. GB $\{(24\bar{1}),(\bar{4}\bar{2}1),0\}$ on Cu. The specimen is compressed perpendicularly to the GB plane from the top at a rate of v=0.81 m/s. A dislocation a) in equilibrium t=0 ns, b) before absorption t=0.62 ns, c) after absorption t=0.94 ns and d) after being nucleated and emitted on the other side t=1.2 ns is shown.

It is noticed that before the absorption, the size of the stacking fault separating the two Shockley partials is reduced. This phenomenon is explained by the fact that the stress field around the grain boundary first affects the leading partial [17, 18]. The simulation runs until the first partial dislocation is nucleated on the opposite side of the GB, at approximately the same location as the earlier impact point (Figure 3d) and subsequently moves away from it.

It is important to add that even though MD can provide much information on the atomistic behavior of GBs and dislocations, the time scales that are currently possible are still very different (much shorter) from what can be done experimentally.

CONCLUSIONS

By calculating the LSTs of all the atoms on bicrystals, it is shown that for the minimum energy configuration of a given planar GB, unique repetitive patterns can be identified. For the case of a higher energy configuration of the same GB, the number of atoms with different LSTs is also higher.

For the minimum energy configuration of the bicrystal $\{(24\bar{1}),(\bar{4}\bar{2}\bar{1}),0\}$ ($\Sigma 7$) in copper, the distance between the two partials of a single edge dislocation approaching the GB gets shorter immediately before being absorbed. It is also noticed that the first nucleation event happens in the same region were the initial dislocation was absorbed.

REFERENCES

1. D. Wolf and S. Yip. Materials interfaces: atomic-level structure and properties. Chapman and Hall (1992).

2. S. M. Foiles, D. Olmsted, and E. A. Holm. Using atomistic simulations to inform mesoscale simulations of microstructural evolution. Proceedings of the Fourth International Conference on Multiscale Materials Modeling, 362 (2008).
3. D. E. Spearot. Atomistic Calculations of Nanoscale Interface Behavior in FCC Metals. PhD thesis, Georgia Institute of Technology (2005).
4. J. D. Rittner and D. N. Seidman. Phys. Rev. B, 54, 6999 (1996).
5. H. Van Swygenhoven and A. Caro. Appl. Phys. Lett., 71, 1652 (1997).
6. P. M. Derlet and H. Van Swygenhoven. Phys. Rev. B, 67, 014202 (2003).
7. Alfthan, S. von, Kaski, K. and Sutton, A. P., Phys. Rev. B, 74, 134101 (2006).
8. B. J. Thijsse. Camelion. Virtual Materials Laboratory, Department of Materials Science and Engineering, Delft University of Technology, The Netherlands.
9. D. E. Spearot, L. Capolungo, J. Qu, and M. Cherkaoui. Comput. Mater. Sci., 42, 57 (2008).
10. Y. Mishin, D. Farkas, M. J. Mehl, and D. A. Papaconstantopoulos. Interatomic potentials for Al and Ni from experimental data and ab initio calculations. In Mat. Res. Soc. Symp. Proc., 538, 535 (1999).
11. Y. Mishin, M. J. Mehl, D. A. Papaconstantopoulos, A. F. Voter and J. D. Kress, Phys. Rev. B, 63, 224106 (2001).
12. P. J. Steinhardt, D. R. Nelson and M. Ronchetti. Phys. Rev. B, 28, 784 (1983).
13. I. Lazic, P. Klaver and B. J. Thijsse. Phys. Rev. B, 81, 045410 (2010).
14. W. Lechner and C. Dellago. J. Chem. Phys., 129, 114707 (2008).
15. J. Li, Modelling Simul. Mater. Sci. Eng., 11, 173 (2003).
16. D. Hull and D. J. Bacon. Introduction to dislocations. Butterworth Heinemann, fourth edition (2002).
17. A. G. Frøseth, P. M. Derlet, and H. Van Swygenhoven, Act. Mat., 52, 5863 (2004).
18. T. S. Byun, Act. Mat., 51, 3063 (2003).

Mater. Res. Soc. Symp. Proc. Vol. 1224 © 2010 Materials Research Society   1224-GG06-04

Application of a 3D-Continuum Theory of Dislocations to a Problem of Constrained Plastic Flow: Microbending of a Thin Film

Stefan Sandfeld[1,3], Dr. Thomas Hochrainer[1,2] and Prof. Michael Zaiser[3]

[1] Institut für Zuverlässigkeit von Bauteilen und Systemen, Universität Karlsruhe (TH), Kaiserstr.12, 76131 Karlsruhe, Germany
[2] Fraunhofer-Institut für Werkstoffmechanik IWM, Wöhlerstr. 11, 79108 Freiburg, Germany
[3] The University of Edinburgh, Institute for Materials and Processes, The Kings Buildings, Sanderson Building, Edinburgh EH9 3JL, United Kingdom

ABSTRACT

The advancing miniaturisation of e.g. microelectronic devices leads to an increasing interest in physically motivated continuum theories of plasticity in small volumes. Such theories need to be based on the averaged dynamics of dislocations. Preserving the line-like character of these defects, however, posed serious problems for the development of dislocation-based continuum theories, while continuum theories based on scalar dislocation densities necessarily stay on a phenomenological level. Within this work we apply a dislocation-based continuum theory, which is based on a physically meaningful averaging of dislocation lines, to the benchmark problem of bending of a free-standing thin film.

INTRODUCTION

Already about half a century ago Kröner [1], Nye [2], Bilby and co-workers [3] and Kondo [4] introduced independently continuum theories of dislocations. Those were based – with slightly different formulations and accents – on a dislocation density tensor, characterising the dislocation state of a crystal. However, already then it was obvious that the dislocation density tensor can only partially describe the defect state of a crystal. The averaged dislocation density tensor measures the 'geometrically necessary' dislocations. Nonetheless, the classical theory is complete in the case when dislocations form smooth bundles of nearly parallel lines with uniform line orientation (as e.g. in [5],[6]).

Hochrainer [7] recently introduced an Extended Continuum Theory (abbreviated by ECT), which contains the classical continuum theory as a special case, but which is furthermore applicable to very general dislocation configurations. We will give a brief introduction of this theory before applying it to the case of micro-bending of a thin film; for a more thorough introduction and derivation of the ECT the reader is referred to [7] and [8].

THEORY

The starting point of ECT is to discriminate dislocation line segments by their line orientation and to average over the so-called lifts of the dislocation lines. Since we consider dislocation glide only we can define the lifted curve in a configuration space where each point (r, φ) contains the spatial point r and the orientation defined as the angle φ between the lines' tangent vector and an arbitrary reference vector. To describe the kinematics of single lines within this space we have to introduce the notion of the generalized line direction L and the generalized velocity V, which denote the tangent to the lifted line and the velocity of the lifted line, respectively:

$$L_{(r,\varphi)}=(\cos\varphi, \sin\varphi, k_{(r,\varphi)})=(l_{(\varphi)}, k_{(r,\varphi)}) \quad \text{and} \quad V_{(r,\varphi)}=(v\sin\varphi, -v\cos\varphi, \vartheta_{(r,\varphi)}). \quad (1)$$

In this definition, l is the spatial line direction, v is the modulus of the spatial velocity v which is normal to l and k is the lines' curvature. The subscripts denote the point of evaluation. The third component of the velocity, ϑ, is a rotational velocity which causes a line to move in orientation direction only and thus to change its orientation by rotation. It reads

$$\vartheta_{(r,\varphi)}=-\nabla_L v_{(r,\varphi)}. \quad (2)$$

With the above notations it is possible to define the so-called 'dislocation density tensor of second order' α'' along with its evolution equations. The second order dislocation density tensor is defined through a density function ρ giving the density of dislocations for each line orientation separately, together with the average curvature k comprised in an averaged lifted line direction as defined in eq. (1). We only state the resulting equations, for details the reader is referred to [7] and [8]. The evolution equation reads

$$\partial_t \alpha''_{(r,\varphi)} =-\text{curl}\left(V_{(r,\varphi)}\times\alpha''_{(r,\varphi)}\right), \quad \text{with} \quad \alpha''_{(r,\varphi)}=\rho_{(r,\varphi)}L_{(r,\varphi)}\otimes b. \quad (3)$$

where $\otimes$ denotes the outer tensor product, $\times$ denotes the cross product, b is the Burgers' vector. The evolution equation for $\alpha''_{(r,\varphi)}$ can be replaced by two evolution equations for the scalar dislocation density ρ and the mean curvature k:

$$\partial_t \rho =-\text{div}(\rho v)-\partial_\varphi(\rho\vartheta)+\rho v k \quad (4)$$

$$\partial_t k =-vk^2+L(\vartheta)-V(k), \quad (5)$$

where we dropped the subscript (r,φ) for brevity. In eq.(4), the first two terms govern the transport of scalar density in the configuration space, while the last term is a source term and accounts for the change of density due to the expansion or shrinkage of loops. The change of curvature during expansion is considered in the first term of eq.(5). The second term in eq.(5) accounts for the change of curvature during rotation of line segments, whereas the last term considers the change of curvature in generalized direction of motion.

In the following we investigate the problem of micro-bending of a free-standing thin film of thickness h. The dimensions of the film in the directions perpendicular to n_s (normal vector of the free surfaces) are assumed large so that the system can be considered homogeneous in these directions. The film is deformed by bending around an axis parallel to the y direction. Dislocations can move on two slip systems that are symmetrically inclined w.r.t. n_s as shown in Figure 1. The Burgers vector of each slip system is perpendicular to the y axis, defining a plane-strain geometry. Since both slip systems are equivalent, it is sufficient to analyse the evolution of the dislocation system for only one of them. Furthermore, we may restrict our investigation to the evolu-

144

tion of strain and dislocation density on a single slip plane as shown in Figure 2 since the system is homogeneous in the directions parallel to the film. The Burgers vector of the considered slip system points in the positive x direction.

Figure 1: *Slip geometry for the symmetrical double slip geometry under consideration.*

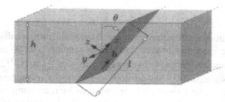

Figure 2: *Definition of the coordinate system for one of the slip systems.*

The shear stress in the slip system can be formally envisaged as the sum of a reference shear stress τ_0 which describes the stress state in a film that is bent to the same radius of curvature R in a purely elastic manner, and a shear stress τ_1 related to the plastic strain γ, the sum of the plastic strains from both slip systems: plastic deformation causes a stress reduction τ_1 proportional to the axial strain, which was obtained assuming isotropic material properties and making the standard assumption that straight specimen cross sections remain straight during bending. The stresses are given by

$$\tau_0(x)=\frac{\sin(2\theta)h}{R(1-v)}G\frac{x}{l} \quad \text{and} \quad \tau_1(x)=\frac{-\sin^2(2\theta)}{1-v}G\gamma(x), \tag{6}$$

where G is the shear modulus of the material, v is Poisson's number, and l is the film width projected on the x direction (Figure 2). The bending moment M (moment per unit length in the y direction) corresponding to a given bending radius R and plastic deformation state is evaluated from the stresses τ_0 and τ_1 as

$$\frac{M}{h^2}=\frac{4}{\sin(2\theta)}\int_0^{1/2}(\tau_0(u)+\tau_1(u))u \ du, \tag{7}$$

where $u=x/l$ and $\tau_{0,1}(u):=\tau_{0,1}(x=ul)$. The scaled bending moment M/h^2 serves as measure of the characteristic stress in our bending simulations.

145

We use a linear-viscous model of over-damped dislocation motion by assuming that the dislocation velocity depends linearly on the difference between the local shear stress τ and the local yield stress τ_y, for the yield stress we assume a Taylor relationship:

$$v = \begin{cases} b/B(\tau-\tau_y) & \text{if } \tau>+\tau_y \\ b/B(\tau+\tau_y) & \text{if } \tau<-\tau_y \\ 0 & \text{otherwise,} \end{cases} \quad \text{where} \quad \tau_y \approx 0.4 \cdot Gb\sqrt{\rho(x)}. \tag{8}$$

We allow dislocations of all orientations to enter or leave the film freely, extrapolating the dynamics inside the film across the boundary. In physical terms our boundary conditions imply that we assume that surface sources have no activation stress other than the yield stress that controls near-surface dislocation motion inside the sample.

DISCUSSION

As initial condition for our simulated bending tests, we assume an isotropic dislocation pattern with zero curvature and space-independent total dislocation density $\rho = 10^{13}\,\text{m}^{-2}$, i.e. initially dislocations of all orientations are present with equal density $\rho_{(x,\varphi)} = (10^{13}/2\pi)\,\text{m}^{-2}$. This initial condition describes a statistically homogeneous and isotropic arrangement of straight dislocation lines. We carry out quasi-static simulated bending tests by increasing the load $\tau_0(x)$ in small steps. In physical terms, this corresponds to slowly decreasing an imposed curvature radius on the thin film. After each stress increment, we solve the equations of motion and evaluate the increase of the plastic strain and the concomitant decrease of local stresses and increase in flow stress. Due to these changes in the internal stress state the strain rate gradually decreases towards zero. We trace this relaxation until the maximum strain rate has everywhere dropped below a prescribed low level and then apply the next stress increment. Typical density and curvature patterns are shown in Fig. 3.

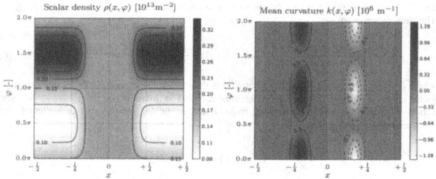

Figure 3: Dislocation density and curvature patterns in the configuration space for a film thickness of $h=3\,\mu m$ and a bending moment (per unit length in y) $M=2*10^{-13}$ GPa m^2

For this bending problem the deformation state is intrinsically heterogeneous and the 'composition' of the dislocation arrangement changes over time due to the growing strain gradients. The dislocation patterns are characterised by the presence of an elastic core region where the stress $\tau_0(x)$ is less than the yield stress corresponding to the initial dislocation density. No deformation activity takes place within this region and, hence, the dislocation density and curvature remain at their initial values. Curvature spatially localizes near the boundaries of the elastic core region which is narrowing with increasing stress. In terms of the orientation coordinate, curvature is strongest for near-screw orientations while edge dislocations are only weakly bent. Density accumulates in the $\varphi = 3\pi/2$ direction, which is the orientation of edge dislocations required to accommodate the bending strain gradient. At the same time, the density of dislocations of other orientations decreases. The decrease is most pronounced for the $\varphi = \pi/2$ orientation, i.e. for edge dislocations of the opposite sign.

The increase of total dislocation density that comes with the accumulation of geometrically necessary dislocations leads, according to the Taylor relation, to an increase in flow stress. As strain gradients are bigger in smaller specimens and therefore the accumulation of geometrically necessary dislocations is more pronounced, this leads to a size dependent hardening. This is illustrated in the left diagram of Fig.4 which shows the scaled bending moment as a function of the average plastic strain $\langle y \rangle = (1/h) \int_{-h/2}^{+h/2} |y(x)| \, dx$. While the initial flow stress is size independent, the hardening rate and the flow stress at finite strain increase with decreasing film thickness. However, the increase of the 'geometrically necessary' density is not only produced by additional dislocations entering through the surfaces, but also by existing dislocations changing their orientation. As a consequence, the increase of the 'geometrically necessary' dislocation density is not accompanied by a proportional increase in total dislocation density (right diagram of Fig.4). Therefore the size effect manifests itself only after an 'incubation strain' during which the geometrically necessary density increase is mainly accommodated by the rotation of existing dislocations. Only after this is exhausted, additional dislocations have to enter from the surfaces and we observe a transition towards a linear increase of dislocation density with strain/strain gradient and a concomitant size dependent hardening as predicted by standard models.

CONCLUSIONS

The evolution of dislocation density in orientation space may be a decisive element in the description of inhomogeneous deformation processes, as changing the dislocation orientation is an important mechanism for providing the 'geometrically necessary' dislocations needed to accommodate strain gradients. Future work will include the treatment of dislocation pile-ups at impenetrable boundaries into the model and comparisons with discrete dislocation dynamic simulations.

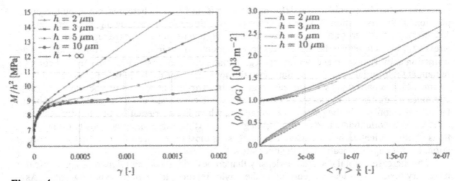

Figure 4:
left: Bending curves for various film thicknesses; the curve for infinite thickness has been calculated by assuming ideal plasticity and using the standard assumption of straight specimen cross-sections.
right: Total and 'geometrically necessary' dislocation density. The upper line group is for $<\rho>$, the lower line group is for $<\rho_G>$

REFERENCES

1. E. KRÖNER, "Kontinuumstheorie der Versetzungen und Eigenspannungen" in Springer, 1958).
2. J. NYE, "Some geometrical relations in dislocated crystals" Acta Metall. 1, 153–162 (1953).
3. B. A. BILBY, R. BULLOUGH, and E. SMITH, " Continuous distributions of dislocations: A new application of the methods of non-Riemannian geometry" in Proc. Roy. Soc. London Ser. A 231, 263–273 (1955).
4. K. KONDO, "On the geometrical and physical foundations of the theory of yielding" in Proc. 2. Japan Nat. Congress of Appl. Mech. 41–47 (1952).
5. R. SEDLÁCEK, J. KRATOCHVÍL, and E. WERNER, "The importance of being curved: bowing dislocations in a continuum description", Phil. Mag. 83, 3735–3752 (2003).
6. A. ACHARYA, "A model of crystal plasticity based on the theory of continuously distributed dislocations", J. Mech. and Phys. Solids 49 (4), 761–784 (2001).
7. T. HOCHRAINER, M. ZAISER, and P. GUMBSCH, "A three-dimensional continuum theory of dislocation systems: kinematics and mean-field formulation" Phil. Mag. 87 (8-9), 1261–1282 (2007).
8. T. HOCHRAINER, "Evolving systems of curved dislocations: Mathematical foundations of a statistical theory", PhD thesis, University of Karlsruhe, IZBS, 2006; Shaker Verlag, Aachen 2007.

Mater. Res. Soc. Symp. Proc. Vol. 1224 © 2010 Materials Research Society   1224-FF05-14

Transition Pathway Analysis of Homogeneous Dislocation Nucleation in a Perfect Silicon Crystal

Hasan A. Saeed , Satoshi Izumi , Shotara Hara and Shinsuke Sakai
Department of Mechanical Engineering, The University of Tokyo,
7-3-1 Hongo, Bunkyo, Tokyo 113-8654, Japan.

ABSTRACT

Transition pathway sampling was carried out for homogeneous dislocation nucleation in perfect crystal Si. The sampling algorithm employed was Nudged Elastic Band method. Results obtained were compared with corresponding results for Cu. The stress and activation barrier ranges were found to be much higher for Si than those reported for Cu. The results also showed that while for lower values of resolved shear stress the dislocation embryo approaches that of a perfect dislocation, for higher resolved shear stress values the embryo is far from perfect. That is, the shear displacement of most particles is considerably less than the Burger's vector. This investigation also demonstrated for the first time that Athermal shear stress for homogeneous dislocation nucleation in Si does not exist, as the crystal undergoes twinning at such high stresses.

INTRODUCTION

Dislocations are responsible for crystalline solids' fundamental mechanical properties of ductility and strength. Dislocation nucleation is a classical example of stress mediated thermally activated transitions. Our interest in investigation of homogeneous dislocation nucleation is because of two reasons: Firstly, homogeneous nucleation, while very closely linked to the ideal strength of a material, is amenable to comparison with dislocation theory, which is not the case with ideal strength. Secondly, investigation of homogeneous nucleation makes it possible to isolate the problem to manageable proportions because of the absence of interfaces and surfaces, which introduce a number of variables into an already complex situation.

Traditionally, most of the atomistic level simulation work done on dislocations has been on mobility, although recently some atomistic work on heterogeneous dislocation nucleation has also been published. Izumi et al., for example, have worked on dislocation nucleation from a sharp corner [1] in Si. In the area of homogeneous dislocation nucleation, the only contribution to date has been Boyer's work [2] on perfect crystal Cu.

In this study we focus on dislocation nucleation in perfect crystal Si. Due to time-scale constraints inherent in dealing with thermally activated phenomena using dynamic atomistic methods such as Molecular Dynamics (MD), we employ the Nudged Elastic Band (NEB) method.

THEORY

Minimum energy paths and saddle point configurations corresponding to a range of shear stress values were determined using the NEB method. The simulation cell consisted of 302,400 atoms, with periodic boundary condition imposed in all directions. The inter-atomic potential employed was SW. The cell dimensions were 23.0×11.5×22.8 nm³.

Shear stresses were applied on the $(111)[01\bar{1}]$ Si slip system by deforming the simulation cell. Due to the non-linearity effects, some unwanted stresses resulted on other planes as well. The effects of those stresses were not determined.

Techniques such as climbing image NEB (CINEB) [3] and free-end NEB (FENEB) [4] were utilized to calculate the activation barrier corresponding to each stress value. The potential ·force vertical to the path on each replica becoming less than 0.005 eV/Å was taken as the criterion of convergence.

Shuffle set dislocation loops were artificially introduced into the crystal. Owing to the importance of the quality of initial input images for successful convergence, these images were chosen from snapshots taken from an unloading MD simulation that resulted in progressively diminishing loops.

DISCUSSION

The dependence of activation energy on resolved shear stress is shown in Figure 1. Unlike the case of homogeneous nucleation in Cu, saddle point configurations corresponding to activation energies less than 1.5eV could not be determined. For higher stresses associated with covalent bonded Si, the dominance of other competing mechanisms such as twinning is likely to be the cause. It can be concluded that for stress values below 6.5 GPa, the activation barrier for dislocation nucleation is lower than that for twinning. For stress values greater than 6.5 GPa however, the situation reverses. Thus there is no Athermal stress for homogeneous dislocation nucleation in Si, hence it is impossible to fit the results into the equation $\Delta E = A \, (1 - \tau/\tau_{ath})^{n}$.

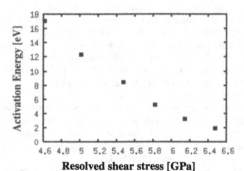

Resolved shear stress [GPa]

Figure 1. Shear stress dependence of activation energy for homogeneous dislocation nucleation in Si. Activation energy values for stresses beyond 6.5 GPa were indeterminate.

Because of severe time-scale limitations, the academically realistic and important region (< 10 eV) associated with low strain rates is not accessible to MD. It is only by utilizing reaction pathway sampling techniques such as NEB that this region becomes accessible. However, in the present case, neither MD nor NEB can handle the region in the vicinity of the Athermal threshold, because such a threshold does not exist for this problem.

Activation energy

Figure 2 compares the stress dependence of activation energy for homogeneous dislocation nucleation in Si and Cu [2]. As expected, homogeneous nucleation in Si is much more stress intensive than is the case with Cu. In fact the stress range is completely different from that in Cu. This is due fundamentally to the difference between metallic bonding and the stronger covalent bonding, as the dislocation width in Si is much narrower than that of Cu, whereas the Peierls stress for Si is much larger than that of Cu.

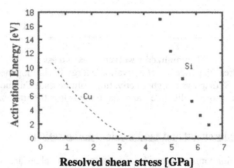

Figure 2. Comparison of stress dependence of activation energy for homogeneous dislocation nucleation in Si and Cu [2]. The process is much more stress intensive in the case of Si.

Various relevant material properties of Si and Cu (all calculated by SW and Mishin potentials respectively, except [5]) are shown in Table 1. The higher stresses involved in homogeneous dislocation nucleation in Si can be attributed to markedly higher values of unstable stacking fault energy, Burgers vector, and $\gamma_{us} \cdot b^2$ for Si as compared to Cu. The quantitative contribution of each parameter needs to be investigated further. The higher stress range for Si also fits in well with its higher ideal strength.

Table 1. Relevant physical properties of Si and Cu

Material	Unstable SF Energy γ_{us} [J/m^2]	Shear Modulus G [GPa]	Burgers Vector b [Å]	$\gamma_{us} \cdot b^2$ [eV]	Ideal Strength [5] τ_{ideal} [GPa]
Si	0.830	64	3.84	0.76	9.62
Cu	0.160	41	1.47	0.02	2.16

Figure 3 shows the comparison of relationship between activation energy and normalized resolved shear stress for Si and Cu. It is obvious that the curve for Si is much steeper than that of Cu. The steeper curve indicates that homogeneous dislocation nucleation in Si is accompanied by a markedly higher activation volume, which means less thermal uncertainty, than is the case with Cu.

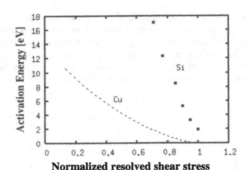

Normalized resolved shear stress

Figure 3. Normalized stress dependence of activation energy for homogeneous dislocation nucleation in Si and Cu. Si displays a higher activation volume than Cu, indicating that the activation process in Si is more 'collective', and that there is less thermal uncertainty involved.

Maximum inelastic displacement and saddle point configuration

Saddle point configurations for a series of stress values are shown in Figure 4. The absolute value of inelastic displacement of atoms is shown in the form of slip vector representation [6]. It is clear from examination of Figure 3 that unlike the case of Cu [2], the dislocation elongates along different Peierls valleys and shows a hexagonal shape even for very small saddle point loop sizes. It is well known that dislocation loops in Si display a hexagonal shape whose edges are parallel to <110> Peierls valleys on the {111} slip plane. This has also been observed in recent low temperature experiments [7] and simulation work [1]. Thus it can be said that our simulation is in good agreement with past work, both experimental and simulation based, on Si. This hexagonal shape can be attributed to the higher Peierls barrier in Si.

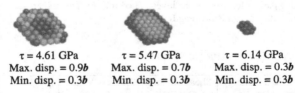

τ = 4.61 GPa	τ = 5.47 GPa	τ = 6.14 GPa
Max. disp. = 0.9b	Max. disp. = 0.7b	Max. disp. = 0.3b
Min. disp. = 0.3b	Min. disp. = 0.3b	Min. disp. = 0.3b

Figure 4. Saddle point configurations corresponding to various stress values. Three regions in terms of applied stress are identified according to the inelastic displacement details.

Figure 5 shows the stress dependence of the maximum inelastic shear displacement of the dislocation embryo as defined by the length of the slip vector normalized by the Burgers vector, for both Si and Cu. In the case of Si, three stress regimes can be identified according to the saddle point configuration and maximum inelastic displacement details (see Figures 4 and 5). In the low stress regime ($\tau < 5$ GPa), the maximum inelastic displacement vector approaches the Burgers vector, and the dislocation embryo approaches that of a perfect dislocation. In this regime, the classical dislocation theory would be effective. However the high activation energy requirements make it highly unlikely for the system to overcome the energy barrier even with the help of thermal activations.

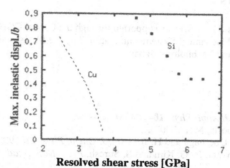

Figure 5. Comparison of maximum inelastic displacement as a fraction of Burgers vector in Si and Cu [2]. Distinct from the case of Cu, with increasing stress the curve becomes increasingly parallel to the x-axis in Si.

In the intermediate stress regime ($5 < \tau < 6$ GPa), the maximum inelastic displacement becomes a decreasing percentage of the length of the Burgers vector. The saddle point configurations in this regime display a diffused core showing in-plane shear perturbations, with the maximum inelastic displacement reaching a lower and lower fraction of the Burgers vector as the stress increases.

It is clear from Fig. 4 that unlike Cu [2], where the maximum inelastic shear displacement approaches zero as the stress approaches the Athermal threshold; in Si the maximum displacement decreases steadily before flattening out. This flattening coincides with the high stress regime ($\tau > 6$ GPa), and in this regime the dislocation core is not diffused. That is, it does not exhibit in-plane shear perturbation exhibited in the intermediate regime. This flattening out might be related to our result that homogeneous dislocation nucleation in Si does not exhibit an Athermal stress. It must be noted that the intermediate and high stress regimes ($\tau > 5$ GPa) are more realistic in terms of the system's ability to overcome the energy barrier (< 10 eV) with thermal activations.

CONCLUSIONS

We report the shear stress dependence of activation energy for homogeneous dislocation nucleation in Si. The higher activation barrier in Si shows that homogeneous dislocation nucleation in Si is much more stress incentive than it is in Cu. We also demonstrate that Athermal shear stress for homogeneous dislocation nucleation in Si does not exist. This is so because at such high stresses, the system undergoes twinning first. This is the first work reported on homogeneous dislocation nucleation in Si.

ACKNOWLEDGMENTS

One of the authors (H. A. S.) was supported through the Global COE Program, "Global Center of Excellence for Mechanical Systems Innovation," by the Ministry of Education, Culture, Sports, Science and Technology, Japan.

REFERENCES

(1) S. Izumi and S. Yip, *J. Appl. Phys.* **104**, 033513 (2008).
(2) R. D. Boyer, Ph.D. thesis, MIT (2007).
(3) G. Henkelman, B. P. Uberuaga and H. Jonsson, *J. Chem. Phys.* **113**, 9901 (2000).
(4) T. Zhu, J. Li, A. Samanta, H. G. Kim and S. Suresh, *Proc. Natl. Acad. Sci. U.S.A.* **104**, 3031 (2007).
(5) S. Ogata, J. Li, N. Hirosaki, Y. Shibutani and S. Yip, *Phys. Rev. B*, **70**, 104104 (2004).
(6) J. A Zimmerman, C. L. Kelchner, P. A. Klein, J. C. Hamilton and S. M. Foiles, *Phys. Rev. Lett.* **87**, 165507 (2001).
(7) J. Rabier and J. L. Demenet, *Scr. Mater.* **45**, 1259 (2001).

Mater. Res. Soc. Symp. Proc. Vol. 1224 © 2010 Materials Research Society   1224-GG05-01

The Vibrational Modes of Model Bulk Metallic Glasses

P. M. Derlet[1*] , R. Maaß[2] and J. F. Löffler[2]
[1]Condensed Matter Theory Group, Paul Scherrer Institut,
5232 PSI-Villigen, Switzerland
[2]Laboratory of Metal Physics and Technology,
Department of Materials, ETH Zurich, 8093 Zurich, Switzerland

ABSTRACT

Bulk metallic glasses exhibit confined low- and high-frequency vibrational properties resulting from the significant bond and topological disorder occurring at the atomic scale. The precise nature of the low-frequency modes and how they are influenced by local atomic structure remains unclear. Using standard harmonic analysis, this study investigates various aspects of the problem by diagonalizing the Hessian of atomistic samples derived from molecular dynamics simulations via a model binary Lennard Jones pair potential.

INTRODUCTION & METHODOLOGY

Bulk metallic glasses (BMGs) exhibit unique vibrational properties. At low frequencies there exists an excess of modes relative to the Debye solid – the Bose peak [1] – which in a three dimensional system may be simply revealed by plotting the low-frequency part of the vibrational density of states (VDOS) divided by the square of the vibrational frequency. The Bose peak regime is considered a fundamentally important material property because its frequency range correlates strongly with the breakdown of transverse linear dispersion. At much higher phonon frequencies there also exists a critical frequency at which there is a transition from extended modes to strongly localized modes – the so-called mobility edge for phonons [2]. Here we investigate some of these aspects within the framework of the Harmonic approximation using atomistic samples that contain both spring-constant and topological disorder.

The BMG samples were generated by quenching from the melt, a 1:1 model A-B binary Lennard-Jones (LJ) [3,4] mixture via constant atom number/pressure/temperature molecular dynamics. We note that the well equilibrated initial liquid state was prepared at a hydrostatic pressure of 14.5 GPa and therefore the quenching procedure involved the simultaneous reduction of both temperature and hydrostatic pressure. Two samples are presented here: sample0 containing 1728 atoms and sample1 containing 13824 atoms. To obtain the final 0K configurations a combination of molecular statics and conjugant gradient relaxation algorithms were used. Fig. 1 displays the volume (per atom), total instantaneous pressure and energy (per atom) as a function of instantaneous temperature during the quench. Inspection of this figure shows that with decreasing quenching rate and/or increasing sample size, the transition from liquid to glass becomes more abrupt with respect to temperature, indicating the underlying first-order phase transition. This trend is confirmed when using quenching rates several orders of magnitude slower, and larger samples sizes – indicating that in smaller samples inherent finite size effects play a role in smoothening out the phase transition. When considering the high quench rates displayed in Fig. 1 it is worth noting that for the present LJ parameterization the employed MD time-step is 0.01 fs.

* Corresponding Author: peter.derlet@psi.ch

Fig. 2a displays the resulting 0K structures and Fig. 2b displays the corresponding pair-distribution functions. An observed insensitivity of the pair distribution functions with respect to quench rate is evident and is also seen for even slower quench rates (not shown). The obtained pair distribution functions agree well with past work [4]. A similar insensitivity is seen when one constructs histograms of the local cohesive energy and stress tensor. These results demonstrate the well-known result that local structural probes characterized by an energy or its first derivative, so successful in identifying defected regions in crystalline structures, are of little use in characterizing amorphous systems.

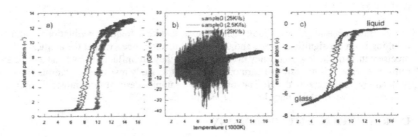

Fig. 1: a) Volume per atom; b) pressure; and c) energy per atom during the quenching of sample0 and sample1 at 25K/fs. For sample0, similar results are shown for an order of magnitude slower quench rate.

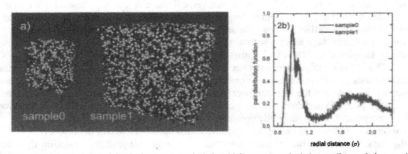

Fig. 2: a) Final BMG samples used for the harmonic analysis in which atoms are shaded according to their type, A and B. b) Corresponding pair distribution functions for the two samples in Fig. 2a.

THEORY AND RESULTS

The $k = 0$ harmonic analysis begins by calculating the Hessian for the system:

$$\Lambda_{ij}^{\mu\nu} = \frac{\partial^2 H}{\partial R_i^\mu \partial R_j^\nu} = \frac{\partial^2}{\partial R_i^\mu \partial R_j^\nu}\left(\frac{1}{2}\sum_{ab} V_{t_a t_b}(R_{ab})\right),$$

where $V_{t_a t_b}(R)$ is the LJ potential between atoms of type t_a and t_b separated by a radial distance R. In the above equation, Latin indices represent atomic labels and Greek indices polarization directions. By solving the eigenvalue equation,

$$\sum_{j\nu}\left(\Lambda_{ij}^{\mu\nu} - \delta_{ij}^{\mu\nu}\omega_n^2\right)u_{n,j}^\nu = 0,$$

the harmonic modes of vibration can be obtained via the eigenfrequencies ω_n and eigenvectors $u_{n,i}^\mu$.

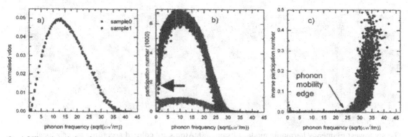

Fig. 3: a) Vibrational density of states for sample0 and sample1. b) Corresponding participation number as a function of phonon frequency. c) Inverse participation number indicating an approximate transition frequency above which phonon states are localized to ~2-3 atoms.

Fig. 3a displays the VDOS for the two samples. The VDOSs display no Van Hove structure demonstrating the strongly disordered nature of the generated BMG structures. The lack of such structure in the VDOS is also typical for systems containing mild degrees of disorder, where similar VDOSs were obtained for atoms within the grain boundary of computer-generated metallic nanocrystalline samples [5,6,7].

Fig. 3b displays the corresponding participation number

$$PN_n = \sum_i \left(\left(u_{n,i}^x\right)^2 + \left(u_{n,i}^y\right)^2 + \left(u_{n,i}^z\right)^2\right)^2,$$

giving quantitative information on the number of atoms involved in each harmonic mode. For a completely delocalized mode, the participation number will equal the number of atoms in the

157

system, whereas for a mode localized to one atom the participation number will equal unity. In. Fig. 3a the large participation numbers corresponding to the three zero-frequency translational modes for sample0 are indicated by an arrow and for sample1 lie outside the range of the graph. At both low and high frequencies there exist modes that involve a reduced number of atoms. Fig. 3c displays the inverse of the participation number. An inspection of Figs. 3b and 3c indicates that the number of atoms associated with the low frequency modes depends on the size of the system, whereas for the high-frequency modes the number of atoms (~2-3) is independent of system size. Establishing a well defined mobility edge in terms of a critical phonon frequency and a corresponding power law, as was done in Ref. [2], was not achieved for these rather small samples.

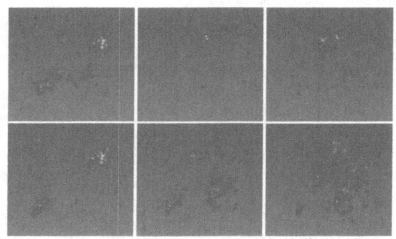

Fig. 4: The six lowest frequency harmonic modes of sample0. Atoms are colored according to their local participation number (dark brown=0.2; bright gold=1). Only atoms with a participation ratio greater than 0.2 are shown. See online document for the color version.

Fig. 4 displays the six lowest-frequency eigenvalues of sample0 by displaying the spatial positions of those atoms whose participation ration is greater than 0.2. Spatially confined modes are clearly evident. Visual inspection of similar low-frequency modes using the same criterion as in Fig. 4 for the larger sample (i.e. sample1) confirmed that the spatial extent of such modes increases with increasing sample size as suggested by the participation number distributions of sample0 and sample1 shown in Fig. 3b.The nature of such low frequency modes remains unclear; past work shows that they may be constructed from a mixture of extended modes and truly localized modes that are independent of system size [8].

An insight into the structural origin of such modes is gained by calculating the (polarization; $\vec{u}_{n,i}$) weighted sum of the local stress tensor of each atom. Fig. 5 demonstrates that those modes below the mobility edge involve atoms that are under tension, whilst those above are in regions of compression. That the high frequency modes correlate strongly with regions of compression within the sample has also been observed recently in metallic grain boundaries [6,7]

– this is a general result as was predicted in ref. [9]. The correspondence of the lower-frequency modes to regions of tension suggests that they might also correspond to regions within the material which are elastically soft, as for instance suggested by recent inelastic x-ray scattering experiments [10]. This aspect will be investigated in future work, in particular how closely related are the tension-to-compression transition frequency and the critical frequency at which the mobility edge for phonons occurs.

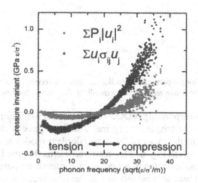

Fig. 5: Participation number-weighted pressure as a function of harmonic mode (phonon frequency).

To investigate the low frequency dispersion a $k \neq 0$ harmonic analysis was performed. Fig. 6 displays the corresponding phonon band structure for sample0 along a particular direction. Here the k points are defined with respect to the simulation super-cell and thus are an artifact of the finite size of the simulation cell – indeed when the sample size increases the corresponding first Brillouin zone would decrease by the same factor and the dispersion bands would undergo zone folding. What is however important here is that linear dispersion is clearly evident for both the two transverse modes and the one longitudinal polarization mode indicating that a Debye solid exists in the long-wavelength limit. A similar result was found irrespective of direction indicating the elastically isotropic nature of amorphous systems in the long wavelength, Debye solid, limit. An inspection of Fig. 6 also reveals numerous low-frequency dispersionless modes similar to those modes visualised in Fig. 4. Importantly, the transverse linear dispersion modes breakup at a frequency corresponding to the onset of such dispersionless modes whereas the longitudinal dispersion mode extends well into this band, appearing not to couple so strongly with these dispersionless modes. The above observations are qualitatively independent of the size of the simulation super-cell used in the calculation.

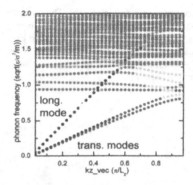

Fig. 6: Phonon dispersion curves along a particular direction are colored according to their polarization type; where blue represents longitudinal, red transverse and green intermediate. See online document for the color version.

Fig. 7 displays the corresponding VDOS (derived from Brillouin zone sampling) divided by the Debye VDOS for this low-frequency regime. The Bose peak is clearly evident precisely at the frequency corresponding to the onset of the dispersionless modes and the break down of the transverse linear dispersion seen in Fig. 6. This result confirms that the Ioffe-Regel limit for the transverse linear dispersion mode occurs at a frequency range corresponding to the onset of the Bose peak [11]. Such a result was also established in the simulation work of ref. [8,12].

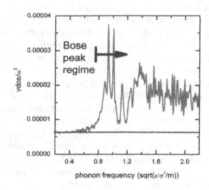

Fig. 7: Low-frequency portion of the VDOS divided by the square of the phonon frequency to indicate the existence of the Bose peak within the harmonic approximation.

160

CONCLUSIONS

Since the transverse linear dispersion mode couples strongly with the low-frequency dispersionless modes, this study suggests that the latter are predominantly transverse in nature. Work is proceeding to gather knowledge of the nature of these low-frequency modes and how they might relate to the structural properties of the bulk metallic glasses.

Investigating these aspects as a function of increasing sample size demonstrates that the dispersionless modes (and the onset of the Bose peak) shift to lower frequencies. Whether or not they converge to a finite frequency has yet to be established.

The longer-term goal is to investigate the connection of such vibrational modes to possible transition pathways for structural relaxation and localized shear deformation zones [13,14].

REFERENCES

1. Phillips,W. A. (ed.) Amorphous Solids: Low-Temperature Properties (Springer, 1981).
2. W. Garber, F. M. Tangerman, P. B. Allen, and J. L. Feldman, Philos. Mag. Lett. 81, 433 (2001).
3. G. Wahnstrom, Phys. Rev. A 44, 3752 (1991).
4. Y. Shi and M.L. Falk, Phys. Rev. B 73, 214201 (2006).
5. P. M. Derlet, R. Meyer, L. J. Lewis, U. Stuhr, and H. Van Swygenhoven, Phys. Rev. Lett. 87, 205501 (2001).
6. P. M. Derlet and H. Van Swygenhoven, Phys. Rev. Lett. 92, 035505 (2004).
7. P. M. Derlet, S Van Petegem, H. Van Swygenhoven, Phil. Mag. 89, 3511 (2009).
8. H. R. Schober and C. Oligschleger, Phys. Rev. B 53, 11469 (1996).
9. A. P. Sutton, Phil. Mag. A 60, 147 (1989).
10. T. Ichitsubo, S. Hosokawa, K. Matsuda, E. Matsubara, N. Nishiyama, S. Tsutsui, and A. Q. R. Baron, Phys. Rev. B 76, 140201(R) (2007).
11. H. Shintani and H. Tanaka, Nature Materials 7, 870 (2008).
12. H. R. Schober, J. Phys.: Condens. Matter 16 S2659 (2004).
13. S. G. Mayr, Phys. Rev. Lett. 97, 195501 (2006).
14. D. Rodney and C. Schuh, Phys. Rev. Lett. 102, 235503 (2009).

Mater. Res. Soc. Symp. Proc. Vol. 1224 © 2010 Materials Research Society   1224-FF11-02

First-principle investigation of electronic structure and mechanical properties of AlMgB$_{14}$

L. F. Wan and S. P. Beckman
Department of Material Science and Engineering, Iowa State University
Ames, IA 50010, U.S.A.

ABSTRACT

The structural and electronic properties of AlMgB$_{14}$ are investigated using *ab initio* methods. The impact of vacancies and electron doping on the crystal's atomic and electronic structure is investigated. It is found that removing metal atoms does not influence the density of states, except for changes to the Fermi energy. The density of states of the off-stoichiometric Al$_{0.75}$Mg$_{0.75}$B$_{14}$ crystal and the AlMgB$_{14}$ crystal with five electrons removed are nearly identical. The removal of six electrons results in an 11% contraction in the crystal's volume. This is associate with the removal of electrons from the B atoms' 2p-states.

INTRODUCTION

There is an immense interest in hard materials for technological applications, such as machine tools[1], grinding medium[2], and wear resistant coatings[3]. The manufacturing industry is a multi-billion dollar per year industry, which means that there is a continued push to increase efficiency, because even small improvements lead to substantial cost savings.[4, 5] Superhard materials, which offer to reduce manufacturing time and increase the lifetime of tools, are naturally an attractive technology.[3, 4, 6]

The boron-rich borides are promising materials for possible use as superhard crystals because they are typically very hard and boron is known to be chemically resistant in situations where diamond is susceptible to chemical attack.[7] The basic unit is the boron icosahedra B$_{12}$, which is strongly bonded other icosahedra and arranged to create a regular lattice of B$_{12}$ units. By counting the bonds within the icosahedra and the number of valence electrons, it is found that an additional two electrons are needed to satisfy the bonds in the B$_{12}$ structure. These electrons typically come from additional atoms that are located between the icosahedra.

A promising class of superhard boron-rich borides is based on the AlMgB$_{14}$ crystal that has been synthesized at the Department of Energy, Ames Laboratory.[8, 9] This crystal class is very different from other boron-rich borides because the Al and Mg metal atoms do not share strong covalent bonds with the boron network.[10, 11] The atomic structure is shown in FIG. 1. The unit cell contains 64 atoms and is composed of four B$_{12}$ icosahedra and eight inter-icosahedra B atoms that are each bonded to three B$_{12}$. Between the layers of B are four Al and four Mg atoms. The Ames Lab specimen is found to have a baseline hardness of 32-35 GPa, and with the addition of a small amount of Ti, the hardness is measured to be as large as 46 GPa.

Surprisingly, aside from the Ames Lab group there have been few experimental studies of this system that focus on the mechanical properties. The primary experimental thrust has emphasized the synthesis of other crystals within the same family, including MgB$_{12}$[12], MgB$_{12}$C$_2$[13], Mg$_2$B$_{24}$C[14], Mg$_2$B$_{14}$[15, 16], AlLiB$_{14}$[17], AlBeB$_{14}$[17], and AlNaB$_{14}$[18]. Although this indicates a great deal of flexibility for chemically alloying the crystal, it does not help to build an understanding of the mechanical behavior.

The few existing theoretical studies have focused on the structure of AlMgB$_{14}$ and its mechanical properties, as characterized by the static bulk modulus.[10, 11, 19, 20] There have also been investigations regarding substitutional impurities on the Al and Mg sites.[10, 20] From these it is known that the metal atoms do not have strong covalent bonds with the B network, but instead contribute their valence electrons to the lattice.[10, 20] It is also known that the off-stoichiometric structure, Al$_{0.75}$Mg$_{0.75}$B$_{14}$, is energetically preferred to the fully occupied AlMgB$_{14}$,[10, 11, 19, 20] an observation that is in agreement with experimental evidence[17]. The bulk modulus has been calculated to be around 200 GPa, and substitutional changes to the metal sites do not result in large changes to this value.[10, 20] The calculated bulk modulus seems slightly lower that what would be expected for a crystal with hardness around 35 GPa.[6, 21]

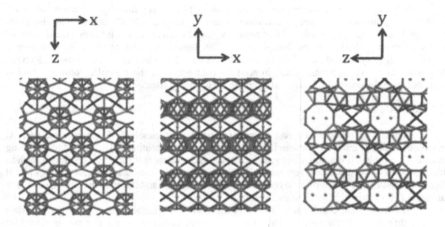

FIG. 1 The crystal structure of AlMgB$_{14}$. The boron atoms are grey, the aluminum are blue, and the magnesium are gold. The B$_{12}$ icosahedra are circled in red and the inter-icosahedra boron atoms are circled in green.

In this paper, *ab initio* total energy calculations are performed to obtain the structural properties and static bulk modulus of AlMgB$_{14}$ and the off-stoichiometric Al$_{0.75}$Mg$_{0.75}$B$_{14}$ crystal. The density of states is calculated and the effect of "electron doping," by changing the total number of electrons, is examined.

METHOD

Density functional theory methods [22-24] are used to investigate the structural, electronic, and dynamical properties of AlMgB$_{14}$ and related crystals. The exchange-correlation interactions are included as local functional of the density, specifically the generalized gradient approximation.[25] In place of the all-electron, r^{-1}, ion potentials, ultrasoft pseudopotentials are used.[26] The wave function is expressed as a truncated plane wave expansion[27, 28] and the Monkhorst-Pack method is used to sample k-points across the Brillouin zone[27, 28]. In this work the cut-off energy is 68 Rydberg and a 4×4×4 k-point grid is used. This is sufficient for the calculated forces to be converged to better than 10 meV/Å. To further verify the convergence for select structures the atomic structure and bulk modulus were calculated for a cut-off energy as large as 85 Rydberg and k-point grid as large as 8×8×8. The increase in the convergence parameters did not result in changes to the structural or elastic properties.

In density functional theory, the charge density is determined by filling the calculated Kohn-Sham states with the valence electrons.[24] In this *aufbau* approach, the addition or subtraction of electrons from the supercell is only a matter of modifying the number of occupied bands. To assure that the calculations do not diverge, due to charging of the supercell, a uniform jellium background is included to maintain charge neutrality. In these calculations the added and subtracted charge is delocalize; therefore, it is not necessary to correct for dipole interactions across supercell boundaries.[29]

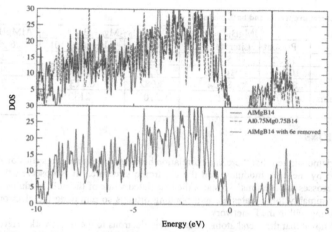

FIG. 2 The top frame shows the density of states for $AlMgB_{14}$ and $Al_{0.75}Mg_{0.75}B_{14}$ crystals. The $Al_{0.75}Mg_{0.75}B_{14}$ DOS matches that of $AlMgB_{14}$ with 5 electrons removed. The bottom frame shows $AlMgB_{14}$ with 6 electrons removed. For both frames the valence band maximum is arbitrarily selected to be the energy zero.

RESULTS AND DISCUSSION

The density of states (DOS) is calculated for the $AlMgB_{14}$ and $Al_{0.75}Mg_{0.75}B_{14}$ crystals and is plotted in the top frame of FIG. 2. The removal of the metal atoms (one each per unit cell) does not effect the shape of the DOS or the band gap, only the position of the Fermi level changes. This suggests that states near the band edges are primarily associated with the B-network and the electrons that are removed come from the metal atoms. To test this hypothesis an $AlMgB_{14}$ crystal is "electron doped" by removing five electrons, which corresponds to the removal of the valence electrons from one Al and one $Mg^{\ddagger}$. It is found that the shape of DOS and position of the Fermi level of the electron doped and off-stoichiometric crystals are nearly identical. This confirms the hypothesis that the B-network determines the rigid band edges, at the gap, and the metal atoms determine the occupation.

The crystal structure and bulk modulus is investigated as a function of electron doping, as shown in Table 1. It is found that electron doping from +2 to -5 electrons does not influence the cell's shape or DOS, only the position of the Fermi level changes. The removal of six electrons still does not influence the shape of the DOS, as shown in the bottom frame of FIG. 2; however, the cell undergoes an 11% volume contraction. There is no obvious explanation for this abrupt change. Because the unit cell is so large, with 64-atoms and 212 electrons, and because the six electrons being removed are delocalized over the entire cell, the doping method is valid. By Löwdin analysis[30] it is determined that the sixth electron removed is likely associated with the B atoms' 2p-states. The impact of this volume change on the mechanical properties is work in progress.

Table 1: Structural properties and bulk modulus

Structure	$AlMgB_{14}$		$Al_{0.75}Mg_{0.75}B_{14}$		$AlMgB_{14}$ missing 6 electron
	Present	Literature[10]	Present	Literature[10]	
a (Å)	5.902	5.895	5.852	5.838	5.700
b (Å)	10.345	10.378	10.261	10.308	9.886
c (Å)	8.112	8.154	8.082	8.113	7.782
Bulk Modulus (GPa)	200	214	210	213	248

CONCLUSIONS

It is demonstrated that there is a correlation between the elastic properties of $AlMgB_{14}$, as characterized by the bulk modulus, and the electronic structure. It is observed that when the Fermi level crosses the B atoms' 2p-states the calculated value of the bulk modulus increases by 50 GPa. Presumably these states are anti-bonding orbitals so removing the electrons allow for stronger bonding within the B-network.

It is known that the metal atoms donate their electrons to the B-network. Here it is shown that modifying the metal atoms results in a shift of the Fermi energy within the density of states. As a result the metal atoms can be used to dope the B-network and modify the electronic properties. It is hopeful that we can use this method to control the mechanical properties of the orthorhombic borides.

$\ddagger$ The Mg pseudopotential used here includes 2p semicore states. By examination of the charge density and Löwdin analysis it is found that these states are highly localized in the ion core and do not contribute to the valence.

REFERENCES

1. M.C. Shaw, *Metal Cutting Principles*. Second ed. 2005, New York: Oxford University Press. 651.
2. I. Finnie, *Some Reflections on the Past and Future of Erosion*. Wear, 1995. **186**(1): p. 1-10.
3. S. Veprek, *The search for novel, superhard materials*. Journal of Vacuum Science & Technology A, 1999. **17**(5): p. 2401-2420.
4. T. Cselle and A. Barimani, *Today's applications and future developments of coatings for drills and rotating cutting tools*. Surface & Coatings Technology, 1995. **77**(1-3): p. 712-718.
5. *Manufacturing in America: A Comprehensive Strategy to Address the Challenges to U.S. Manufacturers*. 2004, U.S. Department of Commerce: Washington D.C.
6. J. Haines, J.M. Leger, and G. Bocquillon, *Synthesis and design of superhard materials*. Annual Review of Materials Research, 2001. **31**: p. 1-23.
7. P. Lazar and R. Podloucky, *Mechanical properties of superhard BC5*. Applied Physics Letters, 2009. **94**(25): p. 251904.
8. B.A. Cook, J.L. Harringa, T.L. Lewis, and A.M. Russell, *A new class of ultra-hard materials based on AlMgB14*. Scripta Materialia, 2000. **42**(6): p. 597-602.
9. T.L. Lewis, *A study of selected properties and applications of AlMgB14 and related composites: Ultra-hard materials*, in *Materials Science and Engineering*. 2001, Iowa State University: Ames.
10. H. Kolpin, D. Music, G. Henkelman, and J.M. Schneider, *Phase stability and elastic properties of XMgB14 studied by ab initio calculations (X = Al, Ge, Si, C, Mg, Sc, Ti, V, Zr, Nb, Ta, Hf)*. Physical Review B, 2008. **78**(5): p. 054122.
11. J.E. Lowther, *Superhard materials*. Physica Status Solidi B-Basic Research, 2000. **217**(1): p. 533-543.
12. V. Adasch, K.U. Hess, T. Ludwig, N. Vojteer, and H. Hillebrecht, *Synthesis and crystal structure of MgB12*. Journal of Solid State Chemistry, 2006. **179**(9): p. 2916-2926.
13. V. Adasch, K.U. Hess, T. Ludwig, N. Vojteer, and H. Hillebrecht, *Synthesis, crystal structure, and properties of two modifications of MgB12C2*. Chemistry-a European Journal, 2007. **13**(12): p. 3450-3458.
14. V. Adasch, K.U. Hess, T. Ludwig, N. Vojteer, and H. Hillebrecht, *Synthesis and crystal structure of Mg2B24C, a new boron-rich boride related to "tetragonal boron I"*. Journal of Solid State Chemistry, 2006. **179**(7): p. 2150-2157.
15. A. Guette, M. Barret, R. Naslain, P. Hagenmuller, L.E. Tergenius, and T. Lundstrom, *Crystal-Structure of Magnesium Heptaboride Mg2b14*. Journal of the Less-Common Metals, 1981. **82**(1-2): p. 325-334.
16. J.S. Peters, J.M. Hill, and A.M. Russell, *Direct reaction synthesis of Mg2B14 from elemental precursors*. Scripta Materialia, 2006. **54**(5): p. 813-816.
17. I. Higashi, M. Kobayashi, S. Okada, K. Hamano, and T. Lundstrom, *Boron-Rich Crystals in Al-M-B (M = Li, Be, Mg) Systems Grown from High-Temperature Aluminum Solutions*. Journal of Crystal Growth, 1993. **128**(1-4): p. 1113-1119.
18. S. Okada, T. Tanaka, A. Sato, T. Shishido, K. Kudou, K. Nakajima, and T. Lundstrom, *Crystal growth and structure refinement of a new higher boride NaAlB14*. Journal of Alloys and Compounds, 2005. **395**(1-2): p. 231-235.

19.   Y. Lee and B.N. Harmon, *First principles calculation of elastic properties of AlMgB14*. Journal of Alloys and Compounds, 2002. **338**(1-2): p. 242-247.

20.   T. Letsoalo and J.E. Lowther, *Systematic trends in boron icosahedral structured materials*. Physica B-Condensed Matter, 2008. **403**(17): p. 2760-2767.

21.   D.M. Teter, *Computational alchemy: The search for new superhard materials*. Mrs Bulletin, 1998. **23**(1): p. 22-27.

22.   P. Hohenberg and W. Kohn, *Inhomogeneous Electron Gas*. Physical Review B, 1964. **136**(3B): p. B864.

23.   W. Kohn and L.J. Sham, *Self-Consistent Equations Including Exchange and Correlation Effects*. Physical Review, 1965. **140**(4A): p. 1133.

24.   R.M. Martin, *Electronic structure : basic theory and practical methods*. 2004, Cambridge; New York: Cambridge University Press.

25.   J.P. Perdew and Y. Wang, *Accurate and Simple Analytic Representation of the Electron-Gas Correlation-Energy*. Physical Review B, 1992. **45**(23): p. 13244-13249.

26.   D. Vanderbilt, *Soft Self-Consistent Pseudopotentials in a Generalized Eigenvalue Formalism*. Physical Review B, 1990. **41**(11): p. 7892-7895.

27.   J. Ihm, A. Zunger, and M.L. Cohen, *Momentum-Space Formalism for the Total Energy of Solids*. Journal of Physics C-Solid State Physics, 1979. **12**(21): p. 4409-4422.

28.   H.J. Monkhorst and J.D. Pack, *Special Points for Brillouin-Zone Integrations*. Physical Review B, 1976. **13**(12): p. 5188-5192.

29.   G. Makov and M.C. Payne, *Periodic Boundary-Conditions in Ab-Initio Calculations*. Physical Review B, 1995. **51**(7): p. 4014-4022.

30.   P.-O. Lowdin, *On the Non-Orthogonality Problem Connected with the Use of Atomic Wave Functions in the Theory of Molecules and Crystals*. The Journal of Chemical Physics, 1950. **18**(3): p. 365-375.

Mater. Res. Soc. Symp. Proc. Vol. 1224 © 2010 Materials Research Society   1224-FF05-31

Effect of alloying elements on the elastic properties of γ-Ni and γ'-Ni$_3$Al from first-principles calculations

Yun-Jiang Wang[1] and Chong-Yu Wang[1,2]
[1]Department of Physics, Tsinghua University, Beijing 100084, China
[2]The International Center for Materials Physics, Chinese Academy of Sciences, Shenyang 110016, China

I. ABSTRACT

The effect of alloying elements Ta, Mo, W, Cr, Re, Ru, Co, and Ir on the elastic properties of both γ-Ni and γ'-Ni$_3$Al is studied by first-principles method. Results for lattice properties, elastic moduli and the ductile/brittle behaviors are all presented. Our calculated values agree well with the existing experimental observations. Results show all the additions decrease the lattice misfit between γ and γ' phases. Different alloying elements are found to have different effect on the elastic moduli of γ-Ni. Whereas all the alloying elements slightly increase the moduli of γ'-Ni$_3$Al expect Co. Both of the two phases are becoming more brittle with alloying elements, but Co is excepted. The electronic structures of γ' phase alloyed with different elements are provided as example to elucidate the different strengthening mechanisms.

II. INTRODUCTION

Nickel-based single-crystal (SC) superalloys are a type of very important materials used in turbine blades for power generation and advanced aircraft engines due to their superior elevated-temperature mechanical properties[1–3]. The superalloys are constituted by precipitate γ' phase (L1$_2$, ordered fcc, Ni$_3$Al based) and matrix γ phase (disordered fcc, solid solution based on Ni). In general, the γ' precipitates play an important role in the mechanical properties of the two-phase γ/γ' alloy. It is well known that the mechanical properties, the lattice misfit, the precipitate morphology and the chemical compositions are all interrelated[4]. The chemical compositions of the constitutive phases controls the sign and the magnitude of the lattice misfit δ, which in turn influences sensitively the precipitate shape and the microstructure evolution during fabrication and application[5, 6]. As a result, the effect of alloying elements on the lattice misfit and elastic properties is critical for a fundamental understanding of the mechanical behaviors of the Ni-base superalloys. Advanced commercial superalloys usually contain many alloying elements. As an example, Re, Ru are the critical additions for the obvious improvement of the performance of alloys. However, to the best of our knowledge, the different strengthening mechanisms related to these alloying elements are not totally understood. Therefore, a deeper understanding of the influence of certain alloying elements on the elastic properties of superalloys is of high interest and should be systematically investigated. In view of the important role of density functional theory (DFT) on studying the mechanical properties of material[7–9], we present here the ab initio investigation of effect of alloying elements on the elastic properties of both γ-Ni and γ'-Ni$_3$Al. The essential alloying elements X (X=Ta, Mo, W, Cr, Re, Ru, Co, Ir) in the fourth generation Ni-base superalloys are all included in our study. Our calculated results are in good agreement with existing experiments[10, 11]. The electronic structures of doped γ'-Ni$_3$Al phase are given as example to elucidate the different alloying mechanisms related to these additions.

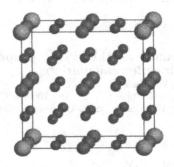

FIG. 1: Supercell of the L1$_2$ Ni$_3$(Al, X) compound, in which the small spheres are Ni atoms, the intermediate spheres Al atoms, and the large spheres alloying element X.

III. CALCULATION METHOD AND MODEL

The first-principles calculations presented here are based on DFT, and have been carried out using the Vienna *ab-initio* simulation package (VASP)[12]. The generalized gradient approximation (GGA) using the Perdew-Wang (PW91) functional[13] of projector augmented wave (PAW) method[14] is adopted for parametrization of the exchange-correlation functional. The cutoff energy of atomic wave functions is set to be 350 eV. A $10 \times 10 \times 10$ k-points in Brillouin zone is adopted with a regular Monkhorst-Pack scheme. The convergence of elastic moduli with respect to the cutoff energy and k-point is tested and gives a satisfactory result. All the internal freedoms of atoms in supercells are fully relaxed. The break condition for the electronic self-consistence and ionic relaxation are 10^{-5} eV and 10^{-4} eV, respectively.

Supercells of $2 \times 2 \times 2$ are adopted for both γ-Ni (or doped with X) and γ'-Ni$_3$(Al, X). Fig. 1 is an example shown for Ni$_3$(Al, X) system. The Al atom at the cube corner is substituted by different alloying elements X to represent the alloyed Ni$_3$(Al, X). Whereas the Ni atom at the Ni supercell corner are replaced by addition for alloyed Ni. The concentration of the alloying elements in our models are 1/32 atom %, which would lead to a obvious strengthening effect. It is should be pointed out that the direct substitution of atom is a rough approximation to investigate the alloying effects. This operation is adopted only for a qualitative estimate of these additions to the elastic properties of both Ni and Ni$_3$Al. If the real alloying effects of these additions are to be investigated, more advanced methods taking the statistical distribution of atoms into consideration are needed in this field.

The theoretical equilibrium volume V$_0$ and bulk modulus B are determined by fitting the total energies versus volume according to the Murnaghan equation of state [15]. The cubic shear constants $C' = \frac{C_{11} - C_{12}}{2}$ and C$_{44}$ are determined by applying an appropriate set of distortions with the distortion parameters varying from δ=-0.06 to 0.06 in the step of 0.01. The applied strain configurations ϵ and the corresponding strain-energy density variations $\Delta E/V_0$ are given as follows: $\epsilon = (\delta, \delta, (1+\delta)^{-2}-1, 0, 0, 0)$ with $\Delta E/V_0 = 6C'\delta^2 + O(\delta^3)$, and $\epsilon = (0, 0, 0, \delta, \delta, \delta)$ with $\Delta E/V_0 = \frac{3}{2}C_{44}\delta^2$.

The elastic constants are calculated using the Hill's averaging method[18]. First, the elastic constants C$_{11}$ and C$_{12}$ are separated from the bulk modulus $B = \frac{1}{3}(C_{11} + 2C_{12})$ and the shear constant C' of a crystal with cubic symmetry. Then, the shear modulus is calculated as the arithmetic Hill average $G = \frac{1}{2}(G_V + G_R)$, where $G_V = \frac{C_{11} - C_{12} + 3C_{44}}{5}$ and $G_R = \frac{5}{4S_{11} - 4S_{12} + 3S_{44}}$ are the Voigt and Reuss bounds, respectively. S_{11}, S_{12}, and S_{44} here

TABLE I: Calculated equilibrium lattice parameters of pure metal γ-Ni, unalloyed γ'-Ni$_3$Al, and lattice misfit δ, in comparison with the experimental values.

	a_γ (Å)	$a_{\gamma'}$ (Å)	δ (%)
Present Calculation	3.515	3.568	1.50
Experiment	3.52[a]	3.57[b]	1.41

[a]Ref. [16]
[b]Ref. [17]

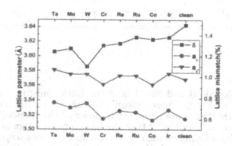

FIG. 2: The lattice parameters of Ni (alloyed with X) (a_γ), alloyed Ni$_3$(Al, X) ($a_{\gamma'}$), and lattice misfit(δ) as a function of the alloying element X.

are the elastic compliances[19]. Finally, the Young's modulus E and Poisson ratio ν are obtained as $E = \frac{9GB}{G+3B}$ and $\nu = \frac{1}{2}(\frac{3B-2G}{3B+G})$, respectively.

IV. RESULTS AND DISCUSSION

A. Variation in lattice properties

The equilibrium lattice parameters of pure metal Ni, pure Ni$_3$Al and the lattice misfit $\delta = \frac{a_{\gamma'}-a_\gamma}{2(a_{\gamma'}+a_\gamma)}$ are listed in Table I. The lattice misfit here is the difference in lattice parameters between doped Ni and doped Ni$_3$Al with the same addition and its concentration. It provides a benchmark for estimating the effect of alloying elements on the lattice properties of Ni-base superalloys. The experimental values are also listed in Table I for comparison. The lattice parameters of Ni and Ni$_3$Al are calculated to be 3.515 Å and 3.568 Å, which is in excellent agreement with the experimental values of 3.52 Å and 3.57 Å, respectively. Moreover, the present calculated lattice misfit is 1.50%, comparable with the experimental result of 1.41%. The consistence confirms the reliability of our calculation.

Furthermore, in order to study the effect of alloying elements on the mechanical properties of Ni-base superalloys, a Ni and an Al atom in the supercell of pure Ni and pure Ni$_3$Al are substituted by a X atom, respectively. The latter one is shown in Fig.1 as an example. The

TABLE II: Calculated elastic constants, bulk modulus, shear modulus, Young's modulus (GPa), G/B ratio, and Poisson ratio of unalloyed Ni_3Al and alloyed $Ni_3(Al, Ta)$ in comparison with the experimental values.

		C_{11}	C_{12}	C_{44}	B	G	E	G/B	ν
Ni_3Al	Present Calculation	229.7	147.5	116.6	174.9	76.8	201	0.44	0.308
	Experiment[a]	227	148	120	174.3	77.0	201	0.44	0.308
$Ni_3(Al, Ta)$	Present Calculation	242.7	151.1	118.9	181.6	81.1	212	0.45	0.306
	Experiment[b]	238	154	130	182	82.8	216	0.45	0.303

[a]Ref. [10]
[b]Ref. [11]

lattice parameters of alloyed Ni and Ni_3Al, and the lattice misfit as a function of alloying elements X are shown in Fig.2, where "clean" means no addition in the systems. The alloying elements X in the figure are arranged in order of the Mendeleev number proposed by Pettifor[20]. It is obvious from Fig. 2 that the lattice parameters of both γ and γ' phases increase with most of the alloying elements, but Co and Cr are excepted. This variation in lattice parameters may correlate with the large atomic radii of alloying elements Ta, Mo, W, Re, Ru and Ir. Whereas the atomic radii of Co and Cr are comparably smaller, as a result the lattice parameters decrease. Fig. 2 also shows that the lattice misfit decreases with all the alloying elements, in which W is the most effective one in decreasing δ. The decreases in δ with alloying element X may influence the microstructure, thereby the mechanical properties of alloys at high temperature[5, 6].

B. Elastic moduli

The elastic moduli of both γ-Ni and γ'-Ni_3Al doped with alloying elements X are all included in our calculations. In Table II, we list the results for Ni_3Al and $Ni_3(Al, Ta)$ as example. The experimental observations are also listed for comparison. Our calculated elastic properties of Ni_3Al and $Ni_3(Al, Ta)$ are in good agreement with the experimental values from Ref.[10] and Ref[11], respectively. Ta considerably increases the elastic moduli of γ' phase, which is consistent with experimental result[11]. This consistence also confirms the validity of our method.

Fig. 3 and Fig. 4 shows our predictions of the elastic properties of alloyed Ni and $Ni_3(Al, X)$ as a function of alloying element, respectively. It is clear from Fig. 3 that different element has different effect on the elastic properties of γ matrix. First of all, all the alloying element except Ta considerably increase C_{11} of Ni. However, their effects on C_{12} behaves differently. Co is the only one that can increase the elastic constant C_{12} of Ni. It is also noticeable that the alloying elements have almost the same effect on C_{44} and C_{11}. Then, Cr, Re and Co consistently increase the bulk modulus of Ni. Interestingly, only Re is effective on increasing the shear modulus of Ni. Finally, all the elements increase the Young's modulus. Generally speaking, Re is the most effect alloying element in increasing the elastic moduli of γ phase. This is consistent with the fact that Re is a crucial alloying element in improving the mechanical performance of Ni-base superalloys[21].

From the elastic properties of alloyed Ni_3Al shown in Fig. 4, we can estimate their

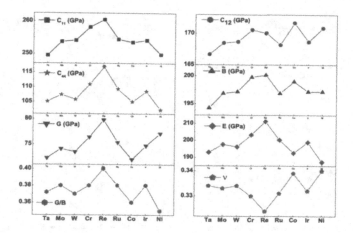

FIG. 3: The calculated elastic constants C_{ij}, bulk modulus B, shear modulus G, Young's modulus E, G/B ratio, and Poisson ratio ν of alloyed Ni as a function of the alloying element X.

different alloying effect on γ' precipitate. The elastic constants C_{11} and C_{12} of Ni_3Al increase with all the alloying elements, in which Re is the most effective one in increasing C_{11}, and Co in C_{12}. While Ru is found to be ineffective on increasing C_{12}. The curve of C_{44} is similar to that of C_{11}, except for Co. The elastic constant C_{44} of $Ni_3(Al, Co)$ is a little smaller than pure Ni_3Al. In comparing elastic properties of various materials, it is sometimes more convenient to deal with the practical moduli such as bulk modulus B, shear modulus G, or Young's modulus E, rather than elastic constans C_{ij}. It is obvious that all the alloying element increasing the bulk modulus of Ni_3Al. Whereas Co has almost no influence on shear and Young's modulus of γ' phase. Similar to γ matrix, Re is also the most effective one on increasing the elastic moduli of γ'-Ni_3Al.

C. Ductile/brittle behavior

According to Pugh's empirical rule[22], the ductile/brittle behaviors of material are closely related to the ratio of G/B and Poisson ratio ν. A high value of G/B and low value of ν is associated with more brittle nature of material. Therefore the increase in G/B or decrease in ν indicates a decrease in ductility. On one hand, all the alloying elements in γ-Ni increase its G/B ratio and decrease its Poisson ratio ν, which is shown in Fig. 3. This variation suggests that all the additions considered in our calculations make Ni more brittle, in which Re is the most obvious one and Co has the most slight effect. As we can predict, a increase in moduli usually accompanies with lose of ductility. This change may be explained by the bonding nature of atoms in the next part. On the other hand, it is also interesting to notice from the G/B and ν curves in Fig. 4 that almost all the X elements can slightly increase G/B ratio and decrease ν value of Ni_3Al. As the circumstance for Ni, Re also has the most effect on the brittle/ductile behavior of Ni_3Al. However, the effect of Co is different. It is the only alloying element that makes an increase in ductility of the Ni_3Al.

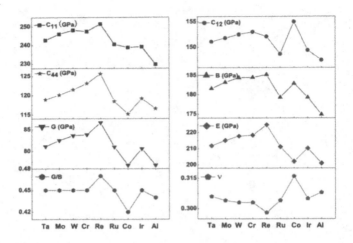

FIG. 4: The calculated elastic constants C_{ij}, bulk modulus B, shear modulus G, Young's modulus E, G/B ratio, and Poisson ratio ν of $Ni_3(Al, X)$ as a function of the alloying element X.

D. Electronic structures of $Ni_3(Al, X)$

In order to develop a deeper understanding of the effect of these alloying elements on the elastic properties of both γ-Ni and γ'-Ni_3Al. Here we provide the electronic structures of the $Ni_3(Al, X)$ systems as an example. Both Local Density of States (LDOS) and charge density difference are included in our calculations. The LDOS curves of the alloying elements X in doped $Ni_3(Al, X)$ systems are plotted in Fig. 5. In contrast with Al, a deep valley separates the bonding and antibonding states of the LDOS for the alloying elements X (Ta, Mo, W, Cr, Re, Ru). The Fermi levels of X (Co, Ru, Ir are excluded) are located in the valleys, indicating those substitutions stabilize $Ni_3(Al, X)$ phase[23]. This feature of LDOS also indicates a strong bonding strength between the alloying element and the host atoms. The bonding is partly covalent. These bonding nature usually suggests a higher elastic moduli[24]. Based on our result, the shear modulus G increases more quickly than the bulk modulus B. As a result G/B increase and leads to more brittleness. This is consistent with the increase of covalent-like bond in the alloyed $Ni_3(Al, X)$ systems. In contrast, there is no obvious bonding and antibonding states of the LDOS of Co, and the Fermi level lies right in the peak of its LDOS. This feature of LDOS is closely related to the metallicity of the alloyed system, which is the reason why only Co addition increases the ductility of γ'-Ni_3Al.

Besides LDOS, the charge density difference on the (100) planes are plotted in Fig. 6. The purpose is to understand the effect of alloying atoms on the bonding properties of the $Ni_3(Al, X)$ systems. The charge density difference of the $Ni_3(Al, X)$ system is defined as $\Delta\rho = \rho[Ni_3(Al, X)] - \rho_{free}[Ni_3(Al, X)] - [\rho(Ni_3Al) - \rho_{free}(Ni_3Al)]$, where ρ_{free} is the superposition of free atom charge density. It can directly reflect the bonding characteristic. From Fig. 6, it can be seen that charge correlation regions due to the electron accumulation appear around the alloying X atoms and their nearest neighbor (NN) Ni atoms. This sharp of charge difference suggests the bonding between X and its NN Ni atoms is stronger than that for Al and its NN Ni atoms in pure Ni_3Al. It indicates enhanced interaction between X and the host atom in $Ni_3(Al, X)$. Except Co, there exists the relative strong bonding that has covalent-like character between X and NN Ni. One can believe that it is this bonding feature of X

174

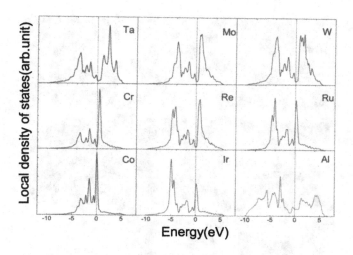

FIG. 5: The electronic Local Density of States of the alloying elements X in the Ni$_3$(Al, X) compounds. The Fermi levels has been shifted to zero.

that leads to more contributions to its elastic moduli. Because the directional distributed covalent-like bond will effectively resist the deformation of a material. It is in turn leading to a high modulus. It is also consistent with the result that the X additions give larger G/B values and decrease the ductility of Ni$_3$(Al, X). The directional distribution of charge different can be found obviously in Ni$_3$(Al, Re) shown in Fig. 6. Further more, we notice that the shape of the charge correlation region of Co in the Ni$_3$(Al, X) system is different from that of the other alloying X atoms. The charge accumulation around Co is spherically symmetric near the atomic site, which is in qualitative agreement with the standard picture of metallic bonding. The feathers of bonding shown in Fig. 6 is closely related to the results that Co increase the ductility of the alloyed system while other additions (X=Ta, Mo, W, Cr, Re, Ru, Ir) make Ni$_3$Al more brittle. Our charge density difference analysis here is consistent with the LDOS.

V. SUMMARY

In summary, the effect of alloying elements on the elastic properties of both γ-Ni and γ'-Ni$_3$Al are systematically investigated by the first-principles method. Results for lattice properties, elastic moduli and the ductile/brittle behaviors are all included in our calculations. The results show that all the additions decrease the lattice misfit, which may affect the mechanical properties of Ni-base superalloys at high temperature. The calculated elastic constants and other moduli of Ni$_3$Al and Ni$_3$(Al, Ta) are in good agreement with the existing experiments. Different alloying elements have different effect on the elastic moduli of γ-Ni. Whereas, all the alloying elements slightly increase the moduli of γ'-Ni$_3$Al expect Co. According to the calculated values of G/B and Poisson ratio, Both of the two phases are becoming more brittle with alloying elements, but Co is excepted. The LDOS and charge density difference of γ' phase alloyed with different elements are provided as example to elucidate the different strengthening mechanisms. The partly covalent-like bond exist be-

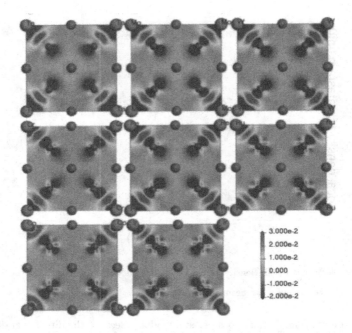

FIG. 6: The charge density difference on the (100) planes of the $Ni_3(Al, X)$ systems and the Ni_3Al system. The small spheres are Ni atoms, the intermediate spheres Al atoms, and the large spheres doped X atoms. Positive (negative) values denote charge accumulation (depletion)[in unit of $e/(a.u.)^3$].

tween alloying element and the host atoms except Co. This is the reason why almost all the alloying elements increase the elastic moduli and brittleness of $Ni_3(Al, X)$, while Co is different. These results may provide useful information for the design of more advanced SC Ni-base superalloys.

VI. ACKNOWLEDGEMENT

A financial grant of the "973 Project" (Ministry of Science and Technology of China, Grant No. 2006CB605102) is gratefully acknowledged.

VII. REFERENCES

[1] T. M. Pollock, and S. Tin, J. Propul. Power **22**, 361 (2006).
[2] C. T. Sims, Superalloy II (Wiley, New York, 1987), pp. 97-131.
[3] Q. Feng, T. K. Nandy, S. Tin, and T. M. Pollock, Acta Mater. **51**, 269 (2003).
[4] R. J. Mitchell, M. Preuss, Metall. Mater. Trans. A **38A**, 615 (2007).
[5] H. Mughrabi, U. Tetzlaff, Adv. Eng. Mater. **2**, 319 (2000).
[6] R. Schmidt, M. Feller-Kniepmeier, Scripta Metall. Mater. **29**, 863 (1993).
[7] M. Šob, M. Friák, D. Legut, J. Fiala and V. Vitek, Mater. Sci. Eng. A **387-389**, 148 (2004).
[8] S. Ogata, Y. Umeno, and M. Kohyama, Modelling Simul. Mater. Eng. **17**, 013001 (2009).
[9] Q. Yao, H. Xing and J. Sun, Appl. Phys. Lett. **89**, 161906 (2006).
[10] M. J. Mehl, B. M. Klein, and D. A. Papaconstantopoulos, in Intermetallic Compounds: Priciples and Practice, edited by J. H. Westbrook and R. L. Fleisher (Wiley, New York, 1994), Vol. 1. pp. 195-209.
[11] R. F. Zhang, S. Veprek, A. S. Argon, Appl. Phys. Lett. **91**, 201914 (2007).
[12] G. Kresse, J. Hafner, Phys. Rev. B **48**, 13115 (1993).
[13] Y. Wang, and J. P. Perdew, Phys. Rev. B **44**, 13298 (1991).
[14] G. Kresse, J. Joubert, Phys. Rev. B **59**, 1758 (1999).
[15] F. D. Murnaghan, Proc. Natl. Acad. Sci. U.S.A. **30**, 244 (1944).
[16] C. Kittle, Introduction to solid state physics. New York: Wiley Intersecience, 1986.
[17] M. H. Yoo, Acta Metall. **35**, 1559 (1987).
[18] R. Hill, Proc. Phys. Soc. London **65**, 349 (1952).
[19] G. Grimvall, Thermophysical properties of materials (NorthHolland, Amsterdam, 1999).
[20] D. G. Pettifor, Solid St. Commun. **51**, 31 (1984).
[21] A. Giamei, D. L. Anton, Met. Trans. A **16**, 1985 (1997).
[22] S. F. Pugh, Philos. Mag. **45**, 823 (1954).
[23] J. H. Xu, T. Oguchi and A. J. Freeman, Phys. Rev. B **36**, 4186 (1987).
[24] Z. M. Sun, R. Ahuja, S. Li, and J. M. Schneiderb, Appl. Phys. Lett. **83**, 899 (2003).

Microcompression & Nanoindentation

Twining and slip activity in magnesium <11-20> single crystal

Gyu Seok Kim [1], Sangbong Yi[2], Yuanding Huang[2] and Erica Lilleodden[1]
[1]Materials Mechanics, Institute of Materials Research, GKSS Research Center,
Max-Planck-Strasse 1, D-21502 Geesthacht, Germany
[2]MagIC-Magnesium Innovation Center,
GKSS Research Center, Max-Planck-Strasse 1, D-21502, Geesthacht, Germany

ABSTRACT

Uniaxial microcompression experiments have been performed on single crystal Mg with a <11-20> compression direction, an orientation unfavorable for basal slip. Results show that the early stages of deformation proceed via both twinning and dislocation plasticity. Twinning leads to a reorientation of the crystal favorable for basal slip, with the <2-1-1-3> aligned with the compression direction. At a critical strain a large strain burst occurs, and is associated with both rapid propagation of the twin and the activation of basal slip within the twin. Such a mechanistic picture of the deformation behavior is revealed through scanning electron microscopy (SEM), electron backscatter diffraction (EBSD) and transmission electron microscopy (TEM) characterization of the deformation structures.

INTRODUCTION

Due to the ever increasing need for energy efficiency, the search for lightweight structural materials is an important and active area of research. Mg, the lightest of all structural metallic materials, shows great promise, although the inherent plastic anisotropy of this hcp metal necessitates a greater basic understanding of its mechanical behavior than is currently known. While many studies have been carried out on Mg single crystal [e.g., 1-3], the difficulty of obtaining truly single crystalline bulk materials complicate an unambiguous study of the twining process. Identification of individual slip mechanisms have been carried out with *in situ* TEM investigations [3], although the stress field is expectedly not uniform, and is made more difficult due to the free surfaces. Therefore, in order to identify slip and twinning activity in Mg, a simpler, known stress state testing technique is needed, while maintaining a deformation volume small enough for comprehensive post mortem characterization. Microcompression testing provides such advantages [4-6]. This relatively new mechanical testing technique has been used to study the mechanical behavior of a variety of materials, with a strong focus on single crystalline fcc metals [5-8]. Just recently this technique has been applied to (0001) oriented Mg single crystal [9, 10], where pyramidal slip was shown to be the governing deformation mechanism. It is therefore of interest to study other orientations of single crystal Mg, where deformation twinning and the activation of other slip systems is possible.

EXPERIMENTAL

A Mg single crystal of orientation (0001) was purchased from Goodfellow, GmbH, with the purity of 99.999%. To obtain the <11-20> orientation from the (0001) disk, the in-plane orientation was identified with EBSD, and a sample with the appropriate surface normal was cut out using spark erosion. After spark erosion, mechanical and chemical polishing was used to remove the severely deformed layer, and EBSD was used to verify the orientation.

Microcompression columns were created from the bulk single crystal using focused ion beam machining on a Nanolab 200 Dualbeam scanning electron (SE) and focused ion beam (FIB) microscope from FEI. Annular cutting was employed with varying probe currents that depend on the final desired geometry. Each column has a nominal diameter of 5µm and a height to diameter aspect ratio of approximately 3:1; no columns with an aspect ratio greater than 4:1 or less than 2:1 were used. Compression tests were performed with a Nanoindenter XP (Agilent) outfitted with a flat punch indenter, using a nominal strain rate of $0.001s^{-1}$ and carried out to varying strain. To evaluate the reproducibility of the experiment, 20 tests have been done. After the microcompression test, cross sectional lamella have been prepared from the columns using FIB milling in order to investigate the twin and slip activity using both electron backscatter diffraction (EBSD, (EDAX, GmbH)) and transmission electron microscopy (TEM, Philips CM200).

RESULTS AND DISCUSSION

The Schmid factors associated with the various slip systems in Mg for uniaxial compression along the <11-20> direction are summarized in Table 1. It can be seen that prismatic slip, pyramidal slip (both π1 and π 2) and tensile twinning all have significant Schmid factors. Only basal slip has zero resolved stress. Since the critical resolved shear stress associated with each slip system is different, we cannot know which deformation modes will be activated *a priori*. Also given in Table 1 are the Schmid factors associated with compression along the <2-1-1-3>, which corresponds to the loading axis for material which underwent tensile twinning. In this case, it is clear that any twinned material is likely to slip on the basal slip system, which is the system easiest to activate in Mg.

Slip System	Schmid factor	
	<11-20>	<2-1-1-3>
Basal:	0	0.44
Prismatic:	0.43	0.12
Pyramidal ($\pi1$):	0.38	0.21
Pyramidal ($\pi2$):	0.44	0.40
Tensile twin:	0.37	0.36

Table 1. Schmid factor of the various slip and twinning systems for Mg for the two relevant loading directions: <11-20> for the parent material, and <2-1-1-3> for the twinned material.

Engineering stress-strain curves, associated with the compression to different stain levels of 6 microcolumns with a nominal diameter of 5µm are shown in Figure 1. The post-compression morphology of these columns is shown in Figure 2, using the same lettering for stress-strain curve and SEM micrograph. It is shown that in three of the columns a critical point occurring at around 1% strain leads to a massive strain burst. In the other three columns, no burst is observed. EBSD and TEM characterization of cross-sections from the column confirm that twinning initiates at the early stage of deformation, prior to the critical point. Twinning is observed in all columns irrespective of whether a critical point was reached, as indicated by the presence of dark bands in post-compression SEM images, as shown in Fig 2 (a-b), or EBSD or TEM analyses. In the columns undergoing massive strain bursts, large slip steps due to basal slip within the twinned region is observed, as shown in Fig. 2 (d-f). While some variation in the initial stress-strain behaviour is observed in Fig. 1, we believe that this is due to statistical differences between columns in terms of pre-existing dislocation content and the ease with which twins can nucleate.

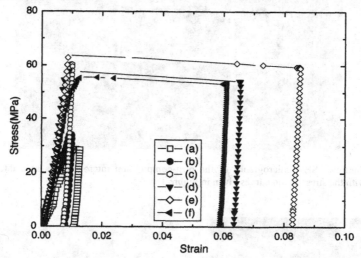

Figure 1. Stress-strain responses for microcompression along the<11-20> direction for 4 columns of mid-height diameter of 5µm. Columns (a)-(c) show plastic deformation with no massive strain bursts, while columns (d)-(f) each show a massive strain burst at around 1% strain.

The initiation and propagation of the twin in column (a) from Figure 2 is shown in Figure 3. The twin initiates at the top of the column with a needle shape, and propagates through the column. Non-basal <a> type dislocations have been observed within the untwinned region, as evidenced by the TEM micrograph in Figure 3(c). The bright field image shows that the Burgers vector is not of the <0001>, or so-called "<c>" type, and therefore is of the <11-20> type, or so-called "<a>" type. Traces of the (0001) planes show that these <a> type dislocations do not lie on the basal plane, and are therefore likely prismatic dislocations. The conclusion that these dislocations are due to prismatic slip is supported by the Schmid factor for prismatic slip given in Table 1. Observations of non-basal <a> dislocations have also been reported by Agnew and coworkers [11].

The orientation of the twin is favorable for basal slip, with the compression axis aligned along the [2-1-1-3] direction, as indicated by the Schmid factor in Table 1. At a critical thickness of the twin, basal slip is able to accommodate a large plastic strain across the entire column diameter. This massive shearing deformation is shown in Figures 4 (a) and 4 (b). A weak beam dark field TEM image is given in Figure 4 (c), showing <c> component dislocations (i.e., either <c> or <c+a> type dislocations) trapped within the twin. This is consistent with the observations of Song and coworkers [12].

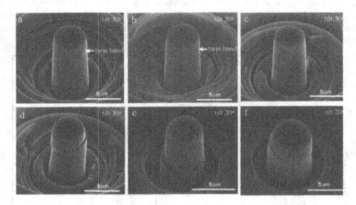

Figure 2. SEM micrographs of the post-compression microcolumns associated with the stress-strain curves given in Figure 1.

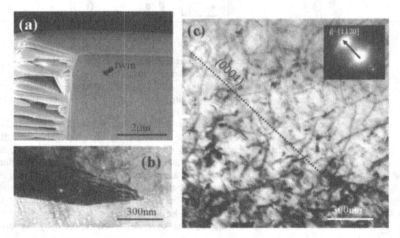

Figure 3. (a) SEM and (b) TEM micrographs of a twin needle initiated within the parent phase associated with column (a) in fig 2. (c) TEM micrograph of non-basal <a> type dislocations within the parent phase. All micrographs are of the early stage deformation structure prior to a massive strain burst.

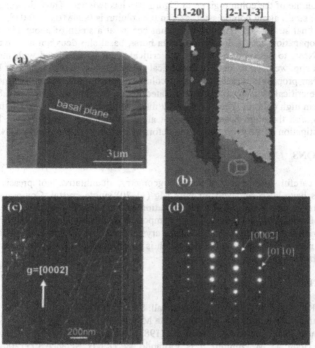

Figure 4. SEM, EBSD and TEM micrographs are associated with the deformation structure after a massive strain burst. (a) SEM micrograph showing a slip step due to basal slip across the entire diameter of the column within the twinned region. (b) EBSD analysis shows that the twin has propagated to nearly the base of the microcolumn. (c) A TEM micrograph and (d) its associated diffraction pattern of the twinned region show that dislocations having <c> component are trapped inside the twin. The zone axis is [2-1-10].

It is not clear, *a priori,* whether the critical point for massive strain burst is due to massive basal slip or to rapid propagation of the twin along the column; post-deformation characterization only tells us that the twin extends through most of the column only after strain bursts, and that large slip steps associated with basal slip within the twin only occur after massive strain bursts. The contribution of the twin to the measured plastic strain along compression axis, ε, is given by:

$$\varepsilon = VM\gamma, \tag{1}$$

where V is volume fraction of the twin, M is the Schmid factor, and γ is the shear strain due to twinning. The twinning shear strain, γ, in magnesium is 0.129(c/a=1.624)[13]. It is shown that

185

80% of the volume of the column given in Figure 4(b) has twinned. Thus, the contribution of the twin to plastic strain along compression axis in the column is 0.048*0.8 = 0.038, or equivalently 3.8% longitudinal strain. Since the critical point occurs at a strain of about 1%, it is clear that rapid twin propagation occurs during the strain burst; basal slip does not account for the entire burst strain. Now, to understand whether the critical point is associated with the activation of massive basal slip, we need to identify the critical twin thickness at which a shear band along the basal plane can propagate across the entire column diameter. A simple geometric argument shows that the critical twin thickness is associated with a longitudinal strain of 1.8%, for a 5μm diameter, 15μm high column. Thus we see that massive basal slip is not possible until a strain of at least 1.8%, and therefore basal slip cannot alone initiate the massive strain burst. A more detailed investigation of the evolution of the deformation structure is ongoing work.

CONCLUSIONS

With careful assessment of sample geometry, quantitative compressive stress-strain measurements have been achieved for pure Mg (11-20) single crystal. Compression normal to the (11-20) shows non-basal slip and the activation of tensile twinning during the early stage of deformation. During twin propagation, <c> component dislocations are trapped within the twin. Tensile twinning leads to a reorientation of the crystal favorable for basal slip. At a critical stress (or strain), a massive strain burst occurs which is associated with further twin propagation and basal slip within the twin.

REFERENCES

1. E.W. Kelley, W.F. Hosford, Trans. Metall. Soc. 242(1968) 5.
2. T.Obara, H.Yoshinga, S.Morozumi, Acta Mater. 21 (1973) 845.
3. A. Couret ,D. Caillard, Acta Mater. 33 (1985) 1447.
4. M. D. Uchic, D. M. Dimiduk, J. N. Florando, W. D. Nix, Science 305 (2004) 986.
5. M. D. Uchic, D. M. Dimiduk, Mat. Sci. Eng. A 400–401 (2005) 268.
6. D.M. Norfleet, D.M. Dimiduk, S.J. Polasik, M.D. Uchic, M.J. Mills, Acta Mater. 56 (2008) 2988.
7. J. R. Greer and W. D. Nix, Phys. Rev. B 73 (2006) 245410.
8. C.V. Volkert and E.T. Lilleodden, Phil. Mag. 86 (2006) 5567.
9. E.T. Lilleodden, Scripta Mater. 62 (2010) 532.
10. C.M. Byer, B. Li, B. Cao, K.T. Ramesh, Scripta Mater. 62 (2010) 536.
11. S.R. Agnew, O. Duygulu, International Journal of Plasticity 21 (2005) 1161.
12. S.G.Song, G.T.Gray III, Acta Materialia 43 (1995) 2338.
13. M.H.Yoo, Met Trans A 12A (1981) 409.

Mater. Res. Soc. Symp. Proc. Vol. 1224 © 2010 Materials Research Society   1224-GG02-05

Indentation crystal plasticity: Experiments and multiscale simulations

Hyung-Jun Chang , Marc Verdier, Marc Fivel
Université de Grenoble, Lab. SIMaP-CNRS, BP 75, St Martin d'Hères, F-38402 Cedex, France

ABSTRACT

This work aims at a quantitative simulation of instrumented indentation test based on physics of crystal plasticity. Indentation loading is associated with a complex 3D deformation path: it can be viewed as an ideal benchmark to test various crystal plasticity assumptions. For large scale indentation (micron size), a 3D numerical simulation using finite element crystal plasticity (FEM) is setup and quantitatively compared to experimental results using critical constraints: the load/stiffness-displacement curves and the surface displacements. Various set of parameters obtained from Dislocation Dynamics (DD) are used. A comparison with experiments shows the dominant effect of initial dislocation density and slip system interactions.

For smaller depth (maximum 100 nm), Dislocation Dynamics coupled simulations to FEM are setup. Since this approach does not provide defect nucleation rules, several strategies are implemented and tested: fitting to Molecular Dynamics (MD) load-depth curve for spherical tip, or automatic generation of deformation accommodating dislocation (GND) for conical tip geometry for example. In this framework, size effects show up in the modification of the dislocation structures with depth through critical expansion of dislocation loops and junction formation.

INTRODUCTION

Mechanics of indentation display a very intense research activity, both in experimental and modeling approaches [1]. With the development of instrumented indentation and scanning probe microscopies, it has become a major tool for small scale mechanical testing (in particular thin film technologies). For metallic single crystal, particularly Cu, extracted mechanical properties from load-depth recordings exhibit several length scale effects (at room temperature, or $T/T_{melting}$ ~1/3 for fcc metals): in the sub-micron depth range (typically sub-200 nm with a sharp apex tip with tip radius around 50 nm, say for a Berkovich tip geometry), the load-depth curves show pop-in phenomena such as displacement bursts in load control mode of operation, along with characteristic 3/2 exponent "elastic Hertzian-like" power law segments [2-3]. In the absence of oxide or brittle coating layers, a required condition for such plastic dynamic instabilities is that the indented volume of the crystal be relatively free of structural defects such as dislocations [4]. Otherwise load-depth curves have a continuous behavior, with the power law exponent evolving towards an expected value of 2 for a conical indenter deforming an elasto-plastic medium. However, this regime which extends to depths of several tens of microns, the hardness as defined by the load divided by the area of contact under load is experimentally not constant (the so called Indentation Size Effect ISE). In terms of defect microstructure accompanying the plastic flow in the above mentioned regimes, one can schematize the first regime by an elasto-plastic transition where plasticity is initiated by dislocation nucleation in a small volume confined by the elastic field of the indenter contact. The second regime covers broadly the expansion of plasticity at larger scales, one observes an evolution of the dislocation microstructure: tangle dislocations walls evolve towards well defined sub-grains boundary at the borders of local crystal rotation domains. This scales with the depth of indentation [5-7].

A single numerical simulation approach covering the whole scale is not yet accessible. We present some recent work here concerning both regimes, with a more specific perspective on the crystalline defect structure evolution. For that matter, dislocation theory based approaches are used. We address some questions such as: how realistic are dislocation based models developed for uniaxial deformation of bulk single crystal applied to indentation ? What is the effect of the initial dislocation density in the single crystal ? What signature from the dislocation interactions can we obtain ? What is the effect of plasticity nucleation criterion on the further stage of plastic flow expansion ? In the first part a continuum constitutive law based on physical crystal micro-plasticity is implemented in FEM and quantitatively compared with experiments. In a second part, various scenarii are tested for initial plastic expansion by setting up a discrete dislocation dynamics simulation.

EXPERIMENTS

Four different high purity Cu single crystals are used: the first three one were cut by spark erosion from large crystal plates (surface normal [111], [002] and [022]) specifically grown and slightly bent for neutron mirror application (ILL, Institut Laue Langevin Grenoble). Their (111) Bragg peak broadening is weak but not negligible and their rocking curve indicates the effect of the slight bending in Table 1. A fourth crystal named B. [8] was grown using a Bridgman process minimizing the initial dislocation density; two surface normals are available for this crystal due to the square section mould used, [111] and [123]. All crystal surfaces are carefully prepared using a Mitchell solution followed by a slow electropolishing in 80% orthophosphoric acid (0.7 V potential). The crystal quality and miscut angle are given in Table 1, [111] ILL and [111]B. crystal differ by their initial dislocation density, the lower being the [111]B. The full width at half maximum (FWHM) along the diffraction vector (θ–2θ) is proportional to the square root of the initial dislocation density (so called micro-strain). A ratio of 5 to 100 in dislocation density between the ILL crystals and B crystal is found. Moreover the initial crystal mosaicity is found in the rocking curves (θ scan) being much larger for crystal [111] ILL than [111]B. In summary, crystal [111]B. is the best crystal in our experiments and specific comparison will be made with [111]ILL crystal when comparing the same surface orientation ((111) plane).

Table I. Surface normal miscut and FWHM of (111) Bragg peaks, Cu radiation.

Orientation surface	Miscut χ (°)	θ (°)	FWHM (111) (θ–2θ °)	FWHM (111) (θ scan °)
[111] ILL	1.3	-0.1	0.06	0.57
[002] ILL	0.8	0.6	0.1	0.56
[022] ILL	7	3	0.32	1.35
[111] B.	0.7	0.01	0.038	0.084

Instrumented indentation tests are carried out to a final depth of 1.4 μm at a constant deformation rate (5×10^{-2} s^{-1}) under the continuous stiffness mode (Nanoindenter MTS-XP). Measurements reported here are made with a sapphire conical tip. The conical geometry is chosen to take into account the deformation anisotropies induced by the different crystal surface symmetry. The final depth is large enough to avoid major effect from the tip defect and ISE (see for example [9-10]). A 3D topographic recording of the tip is measured by tapping mode AFM (Veeco 3100) and the resulting area versus height function of the tip is fitted using a sphero-conical shape function (cone half angle of 71.2°, tip radius 3 μm leading to a 200 nm height of sphere like shape). This geometry is used as well for the numerical simulations.

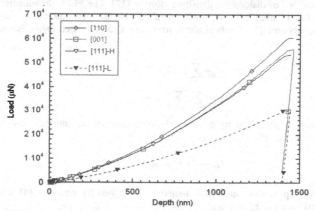

Figure 1. Effect of surface orientation on load depth-curves –[111]-H stands for [111] ILL crystal (high density of dislocation) and [111]-L for [111]B. crystal (low dislocation density).

The load-depth curves for the different crystal are reported in figure 1: one can notice a large variation in load (50%) at full depth between the ILL and B. crystals. The surface displacements around each imprint are recorded by AFM in tapping mode and exhibit the expected symmetry due to the {111} slip systems on each surface normal, i.e. 4 fold symmetry for [001], 3 fold symmetry for [111] and 2 fold symmetry for [011], see for example Fig. 2b and 5b.

SIMULATION AND RESULTS

FEM crystal plasticity

Previous works using FEM classical crystal plasticity (no strain gradient formulation) on indentation of Cu single crystal have been published (see for example [6, 11-14]), but without either the use of physical parameters based constitutive law or the report of quantitative experimental agreement. Our 3D numerical simulations are carried out using Finite Element Method [15] with a dedicated material user subroutine (VUMAT) for crystal plasticity. The simulated crystal is a cylinder (radius 100 μm) meshed by 10959 nodes and 10308 elements along 24 sectors of 15°. The indenter is rigid and has the sphero-conical shape fitted experimentally. Arbitrary Lagrangian and Eulerian (ALE) in explicit integration scheme is used and a quasi-static loading is achieved by careful monitoring the velocity of the indenter to minimize its kinetic energy at surface contact and initial unloading sequences. The imposed velocity (v) of the indenter corresponds to an equivalent strain rate $d\varepsilon/dt$ of 10^{-1} s^{-1} (v/h, h being the displacement) at initial contact and before the unloading and 1 s^{-1} in the intermediate loading regime.

Anisotropic elastic stiffness tensor is used, the three independent coefficients are reported in Table II [16]. Plastic behavior is implemented with a physical model based on Kocks-Mecking constitutive equations of dislocation density evolution [17]. The FEM implementation proposed by Tabourot [18] is used and is briefly detailed. Plastic deformation is summed on all (111)<110> slip systems (1), resolved shear stress are also computed using Schmid tensor $\mathbf{M}^s$, equation (2).

$$\varepsilon_{ij} = \sum_{s=1}^{12} \gamma^s M_{ij}^s \tag{1}$$

$$\tau^s = \sum_{i=1,3} \sum_{j=1,3} \sigma_{ij} M_{ij}^s \tag{2}$$

Plastic strain rate is described by a classical power law taking into account the strain rate sensitivity (m), equation (3).

$$\dot{\gamma}_s = \dot{\gamma}_0 \left(\frac{\tau_s}{\tau_\mu^s} \right)^{\frac{1}{m}} sign(\tau_s) \tag{3}$$

The 12x12 inter-slip system interaction matrix is described by equation (4) where ρ_p is the dislocation density on slip system p :

$$\tau_\mu^s = \mu b \sqrt{\sum_{p=1}^{12} a_{sp} \rho_p} \tag{4}$$

These coefficients a^{sp} are obtained by the forest cutting mechanisms [19], due to symmetry only six independent parameters are required. A constant value for all $a^{sp} = 0.09$ corresponds to Taylor equation or they can be computed using Dislocation Dynamics (DD) simulations [20-22]. The rate equation describing the evolution of dislocation densities ρ_s on each slip system is given equ.(5):

$$\dot{\rho}_s = \frac{1}{b} \left(\frac{\sqrt{\sum_{p=1}^{12} a_{sp} \rho_p}}{K} - 2 y_c \rho_s \right) |\dot{\gamma}_s| \tag{5}$$

which is the core equation to describe the various stage of work hardening rate of fcc single crystal [17]. The first term corresponds to the storage rate of dislocations and has a length dimension: it is a ratio of the dislocation reaction between slip system (junction formation) through a matrix (a), and a semi-phenomenological scaling parameter K describing the self-similarity of the mean free path of a dislocation. The storage term physically corresponds to the maximum rate of work hardening of a single crystal (stage II work hardening) [17]. Recent works using DD simulation [21-22] propose an analytical evaluation for this parameter, but in

190

this study we use a commonly accepted reference value around 30 [17]. The second term of equ.(5) represents the dynamic recovery rate, mainly by pair annihilation through cross-slip mechanisms (y_c, radius of capture depending on the stacking fault energy [17]), b is the Burger's vector. Based on the physical parameters of equation (5), it allows to reproduce most of fcc metals (Al,Cu, Ni...) work hardening behavior [17]. The set of parameters used here are the one for Cu.

In summary, only the initial dislocation density is considered as an adjustable parameter for the constitutive law of deformation, Table II summarizes the set of physical parameters used in this work. To study the effect of the interaction matrix (α^{sp}), two sets can be used corresponding to Taylor model (named *homogenous* case), or a *heterogeneous* case induced by respectively self, coplanar, orthogonal, glissile, Lomer and colinear junction reactions [22,23]. The coefficients of (a) controlling the dislocation rate is also made of 6 independent parameters, two situations are tested here: a set of values proposed by Fivel [20] (called *normal* here), or a set similar to α^{sp} as proposed by Kubin et al. [21] (called *same*). Finally, two values of K (36 – standard [17]) or *high* (100) are tested as well. Therefore a total of 6 different set can be tested, namely Taylor-normal, Taylor-same, Taylor-normal-highK, Heterogeneous-normal, Heterogeneous-same and Heterogeneous-normal-highK.

Table II. Physical parameters input

Elasticity	C_{11}	168.4 GPa
	C_{12}	121.4 GPa
	C_{44}	75.4 GPa
Plasticity	$\dot{\gamma}_0$	$3.7\ 10^{-9}\ s^{-1}$
	m	.01
	y_c	$1.43\ 10^{-9}$ (m)
	K	36 or 100 (*high K*)
	α^{sp}	0.09 (*Taylor*)or (0.122, 0.122, 0.07, 0.137, 0.122, 0.625) (*heterogeneous*)
	a^{sp}	$a^{sp} = \alpha^{sp}$ (*same*) or (0.01, 0.4, 0.4, 0.75,1,0.4) (*normal*)
	b	$2.56\ 10^{-10}$ (m)

The influence of the various crystal surface orientation does not lead to a strong variation of the load-depth curve within the range of initial dislocation density studied here (between $10^{12}\ m^{-2}$ to $10^{14}m^{-2}$) as observed experimentally [24] and in recent simulations [12]. For example in Taylor-normal condition, a 8% variation in load at the same final depth is obtained between the extrema ([011] and [123] orientation) (not shown). A detailed analysis of the contact stiffness and the associated area of contact show excellent agreement with evaluation of indentation modulus computed for high symmetry surface orientation [25] and the experiments [26].

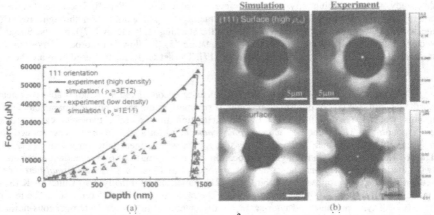

Figure 2. Effect of initial dislocation density (in m^{-2}) on [111] surface normal on (a) simulated load-depth curves (Taylor model),(b) Corresponding surface displacements in simulation and experiments, (Heterogeneous-same set of parameters),scale bar in µm.

More remarkably, a very good agreement is found with the experimental curves using good order of magnitude of the initial dislocation density (Fig. 2a). The effect of initial dislocation density is a major parameter for comparing the model with experiments. It is also remarkable Fig. 2b that the corresponding surface displacements are also in very good agreement with the experiments both in area of contact (not shown) but also on the quantitative profile of displacement for all orientations and crystal conditions studied here. With a higher initial density of dislocation for crystal [111]ILL (therefore a higher equivalent von Mises yield stress criterion) the plastic flow leading to 6 pile-up is much higher (200 nm versus 50 nm height) and closer to the area of contact, Fig.2b.

To get further insight into the effect of the dislocation interaction parameters, Fig. 3 shows the effect of various sets of parameters on both surface displacements and load-depth curves. For an equivalent initial dislocation density on [111] surface normal, the heterogeneous hardening matrix leads to a mechanically much harder response than for the Taylor model. The dislocation junction strength has a weaker effect of on load-depth curves when comparing *same* and *normal* set.

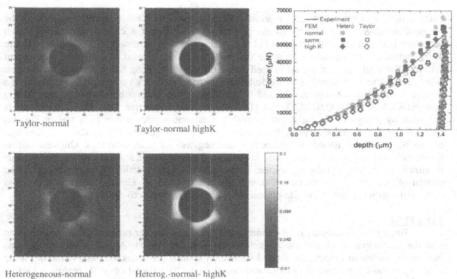

Taylor-normal   Taylor-normal highK

Heterogeneous-normal   Heterog.-normal- highK

Figure 3. Effect of α,a and K parameters, [111] orientation, initial dislocation density per slip system $3.10^{12}m^{-2}$ on surface displacements (scale bar in μm) and load-depth curves

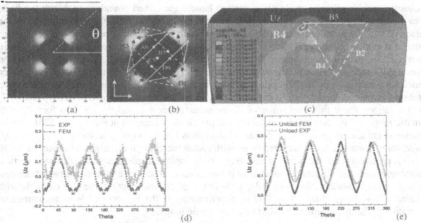

Figure 4. [001] surface displacements (same vertical scale) for (a) Heterog.-same case, (b) experimental with the projection of the slip system flow indicated by arrows (c) cut along B plane ('signed' dislocation density of B4 slip system) and normal displacements (top),(d) –(e) quantitative comparison of surface displacements around the remaining imprint for (d) the profile of contact area, (e) maximum height of pile-up around angle θ defined in (a).

193

Surface displacements show also marked difference between Taylor and heterogeneous: the plastic flow due to slip activity in the pile-up profile displays a clearer separation (trough formation between adjacent pile-up in [111] crystal orientation for example). As shown in figure 4, the upheaval of material forming the different pile-up are related to individual slip system activity. The [001] surface crystal is used here as illustration: using Schmid and Boas notation (see [23]). It can be rationalize in the following manner: Burger's vectors perpendicular to the surface normal are not activated (5 and 6). Each pile-up is a contribution of one pair of slip system,(A2,A3), (B2,B4),(C1,C3) and (D1,D4) and the respective flow of each system is indicated by arrows in Fig.5b. A cut through the maximum pile-up (top right one in Fig. 4a) made along the B plane reveals the 'signed' dislocation density for system B4: the sign convention is positive for out of surface flow and negative for inner solid flow. Outward flow is localized below half of the pile-up and inwards in the solid in the studied case of Fig. 4c. The localized activity on a given slip system leads to local crystal rotation. This matches with the pattern of opposite local crystal rotation domains measured by electron diffraction in cross section of imprints close to the pile-up and below the surface respectively [5-7,27].

DDD-FEM

To gain some insight into the indentation process at smaller range of depth one needs to quit the continuum level and use discrete dislocation description. Although MD simulations include nucleation of defects, the limited simulation volume precludes the observation of plastic zone expansion corresponding to indentation depth in the [0-100 nm] range. Therefore we set up a coupled Dislocation Dynamics simulation [28] with finite element method (CAST3M) [29]. To deal with large density of dislocation in a reasonable amount of CPU time, the dislocation dynamics code is based on edge-screw line discretization. The superposition scheme proposed by v.d.Giessen and Needleman et al.[30] is used to handle the mixed boundary problem: the FEM solves the boundary conditions (free surface and the imposed elastic stress field imposed by the displacement of the indenter) and the DD takes care of the plastic strain response and back stress of the dislocation structure [20,28]. Specific contact algorithms are developed to obtain convergence of displacements under the indenter and quasi-equilibrium state of the dislocation structure (DD code is iteratively run at sub-steps to reach quasi-static configurations) [26]. Since no nucleation rule exist in the DD approach, our aim is to test the response of various criteria and we present some preliminary results corresponding to different assumptions.

MD simulations show that nucleation of dislocations under a round tip leads to formation of prismatic loops below the surface, matching the plastic displacement at the surface [26]. This is the same as nucleating geometric necessary dislocations loops. In our case study, a [111] crystal surface, three prismatic loops are introduced with radius matching the contact area (see Fig.6(a)), therefore producing by the sum of the three overlapping displacement field a net vertical displacement under the indenter. To address the nucleation process we investigate two strategies. The first one is to nucleate loops in order to follow a given load-depth curve (for example with the same behaviour as obtained by MD or experimentally), the second one is to accommodate systematically the displacement by GND-like prismatic loops.

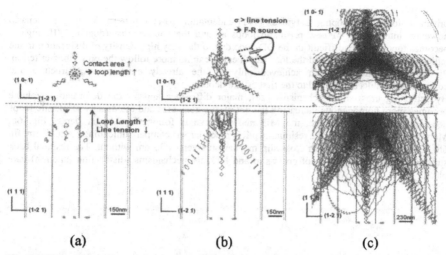

Figure 5. The dislocation microstructure under load for (a) 5 nm, (b) 10 nm and (c) 60 nm of indentation depth, normal and section view. The schematic in top (b) describes the geometry of Frank Read activation from a prismatic loop.

Figures 5 and 6 show results obtained at different depths for a spherical indenter geometry (radius 150 nm, matching most of tip defects) down to a few tens of nm, with the nucleation criteria following a linear (MD-like) force-depth curve and no cross-slip activated.

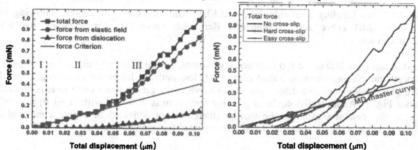

Figure 6. (a) Applied load versus penetration depth from dislocation dynamics corresponding to the microstructures in figure 5, (b) effect of cross-slip loading-unloading curves.

The microstructure shows punch-in prismatic loops into the volume for the first 10 nm (stage I in Fig.6a) followed by the development of glissile dislocations loops at larger depth (stage II, Fig.6a). This corresponds to the matching of the local stress (contact stress and dislocations field) with the radius of the prismatic loop to activate Frank-Read mechanisms from the prismatic loops, developing a typical [111] indentation rosette pattern. With this scenario however, introduction of new prismatic loops around the contact area (domain III, Fig.6a) becomes topologically difficult in the DD code due to the very high density of dislocation in the lattice (simulation artifact) and the load-depth curve can no more follow the prescribed criterion. Plastic displacement is then achieved only by the already existing microstructure and computation results are limited to the first 50 nm depth.

By activating cross-slip mechanisms [28], major difference appear: the deformation volume becomes more homogenous increasing the net plastic zone size and the density of glissile loops decreases. As a consequence, the MD master curve is followed to larger depth, Fig. 6b. Moreover, after unloading, the residual depth is much larger: compare unloading from 60 nm, fig 6b, 50 nm residual depth with cross-slip mechanisms versus 35 nm without. The residual back stress is reduced in the case of cross-slip and locking mechanisms (dislocation junction) take place.

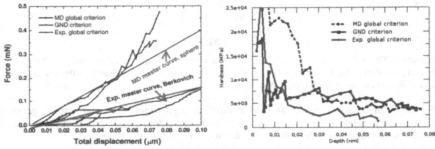

Figure 7. (a) Loading and unloading (10,30,60 nm) of simulated curves following an extrapolated MD curve and experimental Berkovich curves.(b) Corresponding hardness (load/contact area)

Instead of using the MD curve and a spherical geometry, we use a linear fit to the first 100 nm of load-depth curve using a conical indenter into [111] low initial dislocation density single crystal, Fig.8a. The resulting computed hardness versus depth curve gets to value close to experimental behavior Fig.7b. Additionally, coplanar junction formation is observed with [1 −1 0] Burger's vector, a slip system parallel to the surface (therefore not activated in a pure compression loading) and displacing the material laterally as expected for a conical geometry.

SUMMARY AND CONCLUSIONS

In the first part we compare quantitatively experimental indentations with a conical geometry on various Cu single crystals with simulations using physically based crystal plasticity constitutive equations. The constitutive law are developed for bulk single crystal uniaxial loading, taking into account the plastic anisotropy induced by the fcc slip systems. We show a remarkable agreement between the experiments and the numerical simulations, using only one adjustable parameter, the initial dislocation density. To our knowledge, this is the first report of such numerical and experimental agreements. The Taylor model brings the right order of magnitude when comparing the mechanical response (load-depth and stiffness-depth curves) and the boundary conditions (surface displacements field), using the right order of magnitude of dislocation density. By studying the plastic flow around the indenter, the best fitted constitutive law corresponds to the heterogeneous set of slip system hardening and dislocation interaction, in agreement with recent studies of single crystal plasticity using dislocation dynamics [21]. Complementary results using different cone angle and experimental cross-section to access to local crystal misorientation is under progress. Nevertheless, the constitutive law does not include kinematic hardening (back stress) and therefore no length scale parameter. ISE can not therefore be reproduced and the resulting load-depth curves have a constant curvature, in contrast with the experimental curves. Since this effect is experimentally weak at large depth (greater than 1 µm) we restrict our experiments and simulation to this length scale. Our approach lays the foundation for more complex boundary conditions, such as the study of interface effects (grain boundary or film on substrate).

To get some insight into a volume/size effect during indentation, we developed a dislocation dynamics approach, coupled with FEM. Due to the lack of a comprehensive nucleation criterion, we tested several strategies. From our preliminary results, several size effects are observed : the expansion of glissile loops from prismatic punch-in loops at a critical scale; the effect of cross-slip mechanism leading to a more plastic response and the formation of dislocation junctions.

REFERENCES

[1] see for example MRS Proceedings on Nanoindentation or *J.Mater.Res.* **24** (3) (2009).
[2] S. Suresh, T.G.Nieh, B.W. Choi, *Scripta Mater.*,**41** (1999) 951-957.
[3] W.W.Gerberich, W.M.Mook et al., *Jal App. Mech. Trans ASME* **73** (2006), 327-334.
[4] H.Bei, Y.F.Gao, S.Shim,E.P.George, G.M. Pharr, *Phys. Rev.B* **77** (2008) 060103.
[5] J.W. Kysar, Y.X. Gan, T.L. Morse, X.Chen, M.E.Jones, *J. Mech. Phys. Sol.* **55** (2007) 1554-1573.
[6] N. Zaafarani, D. Raabe, R. N. Singh, F. Roters, S. Zaefferer, *Acta Mater.*, **54** (2006), 1863.
[7] M.Rester, C.Motz, R.Pippan *Phil. Mag. Lett.* **88**, 879 (2008).
[8] Crystal kindly provided by Prof. M. Niewczas, McMaster Uni. (Ca).
[9] K.W. McElhaney, J.J. Vlassak, W.D. Nix, *J. Mater. Res.* 13 (1998) 1300-1306.
[10] Y. Liu, A.H.W. Ngan, *Scripta Mater.* **44** (2001) 237-241.
[11] Y. Wang, D. Raabe, C. Kluber, F. Roters, *Acta Mater.*, **52** (2004), 229.
[12] O. Casals, J. Ocenasek, J. Alcala, *Acta Mater.* , **55** (2007), 55.
[13] N. Zaafarani, D. Raabe, F. Roters, S. Zaefferer, *Acta Mater.*, **56** (2008), 31.
[14] Y.Liu, M.Yoshino,H.Lu, R.Komanduri, *Int. J. Plast.* 24 (2008) 1990-2015.
[15] ABAQUS version 6.5, Analysis User's manual, Hibbitt, Karlsson and Sorensen, Pawtucket, RI, 2004.
[16] G. Simmons and H. Wang,Single Crystal Elastic Constants and Calculated Aggregate Properties: A Handbook (2nd ed.). MIT Press, Cambridge, MA, 1971.
[17] UF. Kocks, H. Mecking, *Prog. Mat. Sci.* 48 (2003) 171-273.
[18] L. Tabourot, M. Fivel, E. Rauch, *Mater. Sci. Eng. A*, **234-236** (1997) , 639.

[19] G.Saada, *Acta Met.* 8 (1960) 841.

[20] M. Fivel, *PhD These*, Institut National Polytechnique de Grenoble (1997).

[21] L. Kubin, B. Devincre and T. Hoc, *Acta Mater.*, **56** (2008), 6040.

[22] B. Devincre, L. Kubin and T. Hoc, *Scripta Mater.*, **54** (2006), 741.

[23] J.P. Hirth, J.Lothe, Theory of dislocations, 2nd Ed.,Wiley Interscience, N.Y.:McGraw-Hill (1982).

[24] Y.Y. Lim, M.M. Chaudhri, Phil. Mag. 82, 2071 (2002).

[25] J. J. Vlassak, W. D. Nix, *J. Mech. Phys. Solids*, **42** (1994), 1223.

[26] H.Y. Chang, *These*, Institut National Polytechnique de Grenoble (2009).

[27] M.Rester, C.Motz, R.Pippan, *J. Mater. Res.* **24** (2009) 647-651.

[28] M. Verdier, M. Fivel, I. Groma, *Modelling Simulation Mater. Sci. Eng.,* **6** (1998), 755.

[29] CASTEM Finite Element Method, kindly provided LMS/DMT CEA Saclay (Fr).

[30] E.van-der-Giessen, A.Needleman, Modelling Simulation Mater. Sci. Eng.,3 (1995),689.

Mater. Res. Soc. Symp. Proc. Vol. 1224 © 2010 Materials Research Society   1224-FF05-09

Dislocation Dynamic Simulations of Metal Nanoimprinting

Yunhe Zhang[1], Erik Van der Giessen[2] and Lucia Nicola[1]

[1] Department of Materials Science and Engineering, Delft University of Technology, 2628 CD Delft, The Netherlands

[2] Zernike Institute for Advanced Materials, University of Groningen, 9747 AG Groningen, The Netherlands

ABSTRACT

Simulations of metal nanoimprinting by a rigid template are performed with the aim of finding the optimal conditions to retain imprints in a thin film on a substrate. Specifically, attention is focussed on the interface conditions between film and substrate and on the template shape. Deeper imprints are obtained when the interface between film and substrate is penetrable to dislocation motion. When the protruding contacts of the rigid template are closely spaced the interaction between neighboring contacts gives rise to material pile-ups between imprints.

INTRODUCTION

Metal nanoimprinting is of great technological interest due to its potential applications in miniaturized systems. While the most common technique to achieve nanoimprints in metal is lithography, e.g. [1], imprinting by mechanical indentation of the film has recently been suggested as a promising alternative approach, see e.g. [2]. The objective of this study is to investigate numerically the ability of a metal film on substrate to retain imprints when indented by a rectangular wave pattern. We focus our attention on the nature of the interface between film and substrate, and on the effect of the spacing between protruding contacts. In this respect the size dependence of plastic properties at the sub-micron size scale is expected to cause a non-trivial interaction of the plastic zones underneath the contacts [3].

At the length scale of interest for miniaturized devices, conventional finite element simulations based on classical continuum plasticity fail in predicting localized stresses and deformations. The approach used in this study is 2D discrete dislocation plasticity [4], where plasticity in the metal film is described in terms of the collective motion of discrete dislocations. The discreteness of dislocations, with an evolving density, is the key element for size dependent plasticity, giving rise to a large deviation of submicron-structure behavior from that of bulk metal. In addition, large number of dislocations gliding out the metal free surface leave surface steps that are comparable in size to the depth of the final imprint.

Dislocations are modeled as line singularities in an otherwise isotropic linear elastic medium. Constitutive rules are supplied for the glide of dislocations as well as their generation, annihilation and pinning at point obstacles. The simulations track the evolution of the dislocation structure during loading, unloading and relaxation and provides an accurate description of the final imprinted profile.

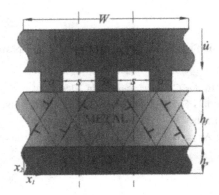

Figure 1. Two dimensional model of a metal thin film on substrate imprinted by a rigid template with rectangular wave profile. Each unit cell contains three flat contacts of size a and spacing s.

MODEL

The thin film is modeled as an infinitely long strip of metal of thickness h_f (see Fig. 1) bonded to an elastic substrate of height h_s. The film is constrained to deform in plane strain under the indentation by a rigid template with a rectangular wave profile. Each protruding flat contact of width a and spacing s is assumed to stick perfectly to the thin film during imprinting. The analysis is performed on a periodic unit cell containing three contacts.

In all simulations the metal film is $h_f = 200$ nm thick and is taken to represent aluminum through its Young's modulus $E = 70$ GPa and Poisson's ratio $\nu = 0.33$. Imprinting occurs at a constant speed $\dot{u} = -2 \times 10^7$ nm/s (corresponding to a uniform strain rate $\dot{\epsilon} = \dot{u}/h_f = -6.7 \times 10^4 s^{-1}$) to a maximum imprinting depth $u_{max} = 10$ nm. Dislocations with a Burgers vector magnitude $b = 0.25$ nm can nucleate from surface sources only (see [5]) with a fixed nucleation time $t_{nuc}=10$ ns. For this nucleation time, the strain rate is sufficiently low to capture all nucleation events and thus correctly describe the yield point. Dislocation glide occurs on two sets of parallel slip planes oriented at $\varphi = 60°$ and $120°$ with the x_1 axis. In the simulations, the active slip planes are spaced at $86b$ and contain a dislocation source just underneath the free surface with nucleation strength $\tau_{nuc} = 50$ MPa, thus the linear density of surface sources is $\rho_{nuc} = 80$ μm^{-1}. Obstacles, which represent forest dislocations in the material prior to loading and small precipitates, are distributed randomly on the slip planes in the metal, amounting to a density of $\rho_{obs} = 30$ μm^{-2}. The critical strength of obstacles is $\tau_{obs} = 150$ MPa. The substrate, of height $h_s = 100$ nm, is treated as being elastic and has the same elastic constants as the film. Two limiting cases are considered for the interface between film and substrate, i.e. the interface is taken to be either impenetrable to dislocation motion or perfectly transparent. In the first case impenetrable obstacles prevent the dislocations from entering the substrate, in the second case dislocations are absorbed in the interface. A fully penetrable interface models the case that the adhesion is so poor that dislocations disappear at the interface without affecting the elastic substrate.

EFFECT OF INTERFACE PENETRABILITY

To study the difference in imprinting when performed with a perfectly penetrable or an impenetrable interface, simulations are carried out for contacts with width $a = 100$ nm spaced 900 nm apart. The imprinting force during loading and unloading is shown in Fig. 2a. As forseeable, the film with impenetrable interface hardens more than that with penetrable interface, since dislocations that pile up at the interface with the substrate obstruct stress relaxation. During retraction of the indenter, the dislocations that have piled up in the film are available to glide in the opposite direction and to partly recover the deformation. This can be seen in Fig. 2b where the surface profiles are shown for the two films after complete unloading. The retained imprints are significantly more evident when the interface is penetrable.   Figure 3 shows the stress state and dislocation structure in the films at

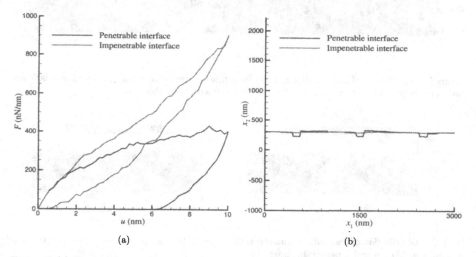

Figure 2. (a) The imprinting force during loading and unloading with an $s = 900$ nm template, and (b) the resulting surface profiles after unloading for different interface conditions. The deformation in x_2 direction is magnified by a factor 10.

$u_{\max} = 10$ nm. As expected, the dislocation density is clearly higher when the interface is impenetrable. The plastic zones underneath the contacts are separated from the neighboring ones in both films. In the following section we will analyze the effect of contact spacing on imprinting.

EFFECT OF TEMPLATE SHAPE

The effect of the template shape is investigated for film-substrate systems with a penetrable interface. To this end simulations are carried out for values of the spacing between

201

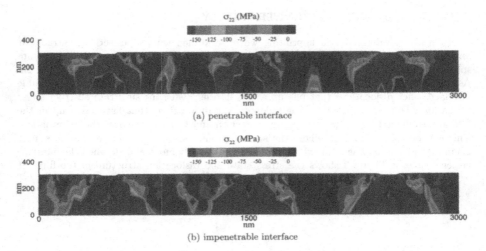

Figure 3. Distribution of dislocations and σ_{22} for a film imprinted by a template with $s = 900$ nm and (a) penetrable interface (c) impenetrable interface with the substrate.

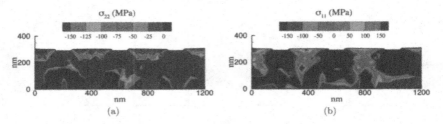

Figure 4. Distribution of dislocations and stress components (a) σ_{22}, (b) σ_{11} for a film with contact spacing $s = 300$ nm and a penetrable interface.

contacts ranging from 300 to 900 nm. Figure 4 shows the stress state and dislocation structure at final imprinting depth for a film indented by a template with $s = 300$ nm. By contrasting Fig. 4 with Fig. 3 one notices that when contacts are close to each other there is more material piling-up between contacts and that the plastic zones overlap.

Figure 5 compares the corresponding surface profile with that for $s = 900$ nm (cf. Fig. 2) as well as $s = 500$ nm. The displacements are magnified by a factor 20 for ease of visualization. The imprints are deeper when the contacts are closely spaced but they are less well defined, i.e. the surface becomes rougher with extrusions that are comparable in size to the imprinting depth. The maximum depth of the imprints relative to the original top surface is 9.6, 7.5 and 7.2 nm for the films with $s = 300$, 500 and 900 nm, respectively.

To quantify the effect of spacing between contacts on the final surface profiles we use the

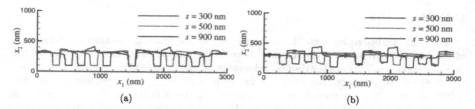

Figure 5. The film surface (a) at maximum imprinting depth; (b) after unloading and relaxation with different s for a penetrable interface. The displacement in x_2 direction has been magnified by a factor 20. Multiple replicas of the unit cell are shown when $s < 900$ nm.

root mean square roughness defined as

$$R_m = \sqrt{\frac{1}{N} \sum_{i=1}^{N} (h_i - h_m)^2}, \tag{1}$$

where N is the total number of nodes on the film surface; h_i is the x_2-coordinate of the ith node of the surface and h_m is the mean height of the top surface. Results in Fig. 6 show that for all template shapes considered here the imprints are better retained when the spacing between contacts is smaller. Moreover, when the spacing is below 500 nm interaction between contacts clearly takes place, material piles up in between contacts during loading and the waviness of the surface is maintained and even increased during unloading (see Fig. 5).

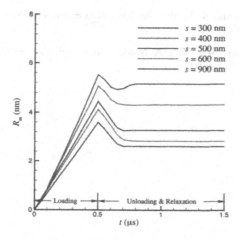

Figure 6. Evolution of the root mean square roughness during imprinting of films with a penetrable interface using different contact spacings s.

203

CONCLUSIONS

Interface conditions between film and substrate are essential in determining the success of the imprinting process: if the interface is perfectly penetrable to dislocation motion the retained surface indents are the deepest achievable. Thus, to the end of creating well-defined nanopatterns an interface that absorbes dislocations is clearly the preferable choice. For a 200 nm thick film, interaction between neighboring contacts occurs when the spacing is below 500 nm for a contact size of 100 nm. When such a spacing is used material pile-ups appear in between contacts and the surface profile becomes more wavy.

ACKNOWLEDGEMENTS

LN is grateful to the Dutch National Scientific Foundation NWO and Dutch Technology Foundation STW for their financial support (VENI grant 08120).

REFERENCES

[1] L. J. Guo, Adv. Mater. **19** (4), 495–513 (2007).

[2] G. L. W. Cross, B. S. O'Connell, H. O. Ozer, and J. B. Pethica, Nano Lett. **7**, 357–362 (2006).

[3] L. Nicola, A. F. Bower, K.-S. Kim, A. Needleman, and E. Van der Giessen, Philos. Mag. **88** (30-32), 3713-3729 (2008).

[4] E. Van der Giessen and A. Needleman, Modelling Simul. Mater. Sci. Eng. **3** (5), 689-735 (1995).

[5] L. Nicola, A.F. Bower, K.S. Kim, A. Needleman, and E. Van der Giessen, J. Mech. Phys. Solids **55**, 1120–1144 (2007).

Mater. Res. Soc. Symp. Proc. Vol. 1224 © 2010 Materials Research Society   1224-FF05-36

On Effective Indenters Used in Nanoindentation Data Analysis

Guanghui Fu[1], Ling Cao[1] and Tiesheng Cao[2]
[1]LC Dental, 43713 Boscell Road,
Fremont, CA 94538, U.S.A.
[2]Department of Ultrasonographic Diagnostics, Fourth Military Medical University,
Xi'an, 710038, CHINA

ABSTRACT

Effective indenter concept was introduced by Pharr and Bolshakov to explain nanoindentation unloading curves. This paper shows that the contact stiffness under a uniform pressure distribution is 57% higher than what is given by the fundamental relation. This is due to the fact that there is no physical indenter that gives a uniform pressure distribution during elastic contact, and the fundamental relation used in nanoindentation data analysis does not apply.

INTRODUCTION

Nanoindentation experiments have become a commonly used technique to investigate mechanical properties of thin films and small volumes of materials. The analysis of the experimental load – displacement ($P - h$) curve is based on the fundamental relation among contact stiffness, contact area and elastic modulus. The slope of the P-h curve, $S = dP/dh$, is defined as contact stiffness and it can be measured from nanoindentation experiments. The fundamental relation relates contact stiffness to the projected contact area (A), Young's modulus of the material (E), and Poisson's ratio of the material (v), as

$$S = \frac{2}{\sqrt{\pi}} \frac{E}{(1-v^2)} \sqrt{A} \qquad (1)$$

The fundamental relation is based on the analytical solution of normal indentation of an elastic half-space by a rigid smooth frictionless axisymmetric indenter [1]. Borodich and Keer [2] prove the fundamental relation by using the indentation shape to relate the load and the depth of the indentation. An alternative approach is to use the contact pressure to relate the load and the indentation depth [3].

The technique of assuming a pressure distribution between an indenter and a half space has been introduced to the nanoindentation data analysis by Pharr and Bolshakov [4] to explain nanoindentation unloading curves. They assume a uniform pressure distribution and use linear elasticity theory to obtain the deformed surface profile. The effective indenter shape is approximated from the deformed surface profile. The mechanical properties of the material are obtained through the effective indenter shape function and the fundamental relation.

There is a lack of studies on the effects of assumed pressure distributions on the fundamental relation. In this paper, we show that a uniform pressure distribution overestimates the contact stiffness by 57%, and it is because there is no physical indenter that gives a uniform pressure distribution during elastic contact.

THEORY

We consider a pressure distribution $p(r,a)$, which applies to a circular area $0 \leq r \leq a$ on the plane boundary of an elastic half-space $z \geq 0$. The problem is considered in the linear theory of elasticity and the half-space is assumed to be isotropic and homogeneous. The stress components have two subscripts corresponding to the appropriate coordinates. E and v are Young's modulus and Poisson's ratio of the half-space. As Fig. 1 shows, the boundary conditions for the half-space at $z = 0$ are

$$\tau_{zr} = \tau_{z\theta} = 0, \, (0 \leq r < \infty) \tag{2}$$

$$\sigma_{zz} = 0, \, (r > a) \tag{3}$$

$$\sigma_{zz} = p(r,a), \, (0 \leq r \leq a) \tag{4}$$

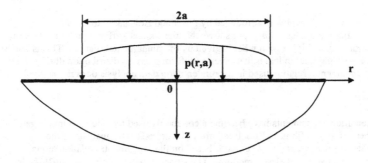

Figure 1. Normal loading on a circular surface area of an elastic half-space

With the uniform pressure distribution, $p(r,a) = q$, the corresponding depth of indentation [5] is

$$h = \frac{2(1-v^2)}{E} \cdot q \cdot a \tag{5}$$

and the total load is

$$P = \pi \cdot q \cdot a^2 \tag{6}$$

From eqn. (5) and eqn. (6), both h and P are functions of a single variable a. We have

$$\frac{dh}{da} = \frac{2(1-v^2)}{E} q \tag{7}$$

and

$$\frac{dP}{da} = 2\pi \cdot q \cdot a \tag{8}$$

The contact stiffness can be rewritten as

$$S = \frac{dP}{dh} = \frac{dP/da}{dh/da} \qquad (9)$$

Noting eqn. (7) and eqn. (8), eqn. (9) becomes

$$S = \frac{\pi E}{1 - v^2} a \qquad (10)$$

or

$$S = \frac{\sqrt{\pi} E}{1 - v^2} \sqrt{A} \qquad (11)$$

Compared to the fundamental relation in eqn. (1), eqn. (11) overestimates the contact stiffness by 57%, a significant error in the nanoindentation data analysis.

DISCUSSION

It is shown in the previous section that a uniform pressure distribution leads to a significant error in the contact stiffness. Indentation problems require that the pressure distribution between the half space and the indenter satisfies the following equation [6]

$$p(r,a) = \int_r^a \frac{g(s)}{\sqrt{s^2 - r^2}} ds \qquad (12)$$

where $g(s)$ is a function of a single variable s, $g(s) \neq 0$ and $0 \leq s \leq a$.

When a uniform pressure distribution is assumed, eqn. (12) becomes

$$q = \int_r^a \frac{g(s)}{\sqrt{s^2 - r^2}} ds \qquad (13)$$

Eqn. (13) is an Abel equation, and we have

$$g(s) = \frac{2}{\pi} q \frac{s}{\sqrt{a^2 - s^2}} \qquad (14)$$

However, eqn. (14) indicates $g(s)$ is a function of two variable s and a, which violates its definition. It shows there is no corresponding indenter to a uniform pressure distribution, and contact stiffness defined by eqn. (1) does not apply to the uniform pressure case.

To maintain a uniform pressure distribution, the shape of the indenter must change with the depth of indentation. This shape change violates the rigid indenter assumption for the fundamental relation.

CONCLUSIONS

In this paper, we show that a uniform pressure distribution leads to significant errors in the fundamental relation, and overestimates the contact stiffness by 57%. This is due to the fact that there is no corresponding indenter to a uniform pressure distribution, and the fundamental relation used in nanoindentation does not apply.

REFERENCES

1. A.C. Fischer-Cripps, *Nanoindentation*, (Springer-Verlag, 2004) New York, New York, USA.
2. F.M. Borodich and L.M. Keer, *Int. J. Solids Strut.* **41**, 2479 (2004).
3. G. Fu and L. Cao, *Mater. Lett.* **62**, 3063 (2008).
4. G.M. Pharr and A. Bolshakov, *J. Mater. Res.* **17**, 2660 (2002).
5. K.L. Johnson, *Contact Mechanics*, (Cambridge University Press, 1985) pp. 107-108.
6. J. Jaeger, *New Solutions in Contact Mechanics*, (WIT Press, 2005), pp. 99.

Mater. Res. Soc. Symp. Proc. Vol. 1224 © 2010 Materials Research Society　　　1224-FF07-08

Similarity Relationships in Creep Contacts and Applications in Nanoindentation Tests

J.H. Lee,[1] C. Zhou,[1] C.J. Su,[1] Y.F. Gao,[1,2] G.M. Pharr[1,3]

[1] Department of Materials Science and Engineering, University of Tennessee, Knoxville, TN 37996, U.S.A.
[2] Computer Science and Mathematics Division, Oak Ridge National Laboratory, Oak Ridge, TN 37831, U.S.A.
[3] Materials Science and Technology Division, Oak Ridge National Laboratory, Oak Ridge, TN 37831, U.S.A.

ABSTRACT

The study of indentation responses of rate-dependent (viscoplastic or creeping) solids has generally focused on the relationship between indentation hardness and an effective strain rate, which can be defined from a similarity transformation of the governing equations. The strain rate sensitivity exponent can be determined from the slope of a log-log plot of the hardness versus effective strain rate, while determining other constitutive parameters requires a knowledge of the relationship between contact size, shape, and indentation depth. In this work, finite element simulations have shown that the effects of non-axisymmetric contact and crystallography are generally negligible. Theoretical predictions agree well with real nanoindentation measurements on amorphous selenium when tested above glass transition temperature, but deviate quite significantly for experiments on high-purity indium, coarse-grained aluminum, and nanocrystalline nickel. Such a discrepancy is likely to result from the transient creep behavior.

INTRODUCTION

The classic Oliver-Pharr approach to determine the elastic modulus and indentation hardness is based on an analysis of rate-independent elastic-plastic contact, for which the correlation between strain hardening characteristics and indentation responses is well established [1-4]. The study of indentation responses of rate-dependent solids has focused mostly on the concept of an effective strain rate, $\dot{\varepsilon}_{eff}$ [5-11]. For instance, $\dot{\varepsilon}_{eff} = \dot{h}/h$ for pyramidal indenter with indentation depth h, and $\dot{\varepsilon}_{eff} = \dot{a}/D$ for a spherical indenter with contact radius a and indenter diameter D.

Consider a J_2 power-law creeping solid characterized by

$$\sigma/\sigma_0 = (\dot{\varepsilon}/\dot{\varepsilon}_0)^{1/m}, \text{ or } \dot{\varepsilon} = A\sigma^m, \qquad (1)$$

where σ_0 and $\dot{\varepsilon}_0$ are the reference stress and reference strain rate, respectively, m is the stress exponent (reciprocal to the strain rate sensitivity exponent), and $A = \dot{\varepsilon}_0/\sigma_0^m$. As explained by Bower et al. [6], at any particular instant, the strain rates and stresses in a pure creeping solid (i.e., no elasticity and no strain hardening) are independent of the history of loading and depend only on the instantaneous velocities and contact radius prescribed on the surface. Consequently, the strain rate and stress fields are identical to those under a rigid flat punch of radius a which indents a creeping half-space at velocity $\dot{h}$. The geometry of the indenter determines the relationship between a and h. Apparently, the effective strain rate should be chosen as $\dot{h}/a$, so that

$$\frac{P}{\pi a^2 \sigma_0} = \left(\frac{\dot{h}}{a\dot{\varepsilon}_0}\right)^{1/m} F_a(m), \tag{2}$$

where P is the indentation load, and $F_a(m)$ depends on the stress exponent m and weakly on the friction condition. For a geometrically self-similar indenter, such as cone with half included angle β (e.g., 70.3° for Berkovich-equivalent cone), we have

$$h = \frac{a}{c(m)\tan\beta}, \tag{3}$$

where $c(m)$ depends on the stress exponent m, and also weakly on the friction condition. The dimensionless parameters, $F_a(m)$ and $c(m)$, can be computed from the punch contact problem, which may be further simplified into a nonlinear elastic contact by replacing the strain rates by strains and velocities by displacements in Eq. (1). Finite element simulations using conical indenters, however, indicate a slight dependence of F_a and c on the included angle [10].

Application of the above theoretical predictions, which are derived from axisymmetric contact, to experimental nanoindentation measurements obtained with a Berkovich triangular pyramid needs to take into account effects of the non-axisymmetric contact shape, the difference between $P/\pi a^2$ and the measured hardness, and the crystallographic dependence caused by anisotropies in slip in single crystals. The effects of non-axisymmetric contact and crystallographic dependence are examined by finite element simulations in this work. Comparisons have been made to experiments for amorphous selenium, annealed indium, coarse-grained aluminum, and nanocrystalline nickel. The first material gives $m \approx 1$ when tested above the glass transition temperature. The rest materials give a variation of stress exponent ~4-10.

CONICAL VERSUS BERKOVICH INDENTATION

Nanoindentation measurement techniques typically use the Berkovich indenter, which is a three-sided pyramid with an equivalent cone angle of $\beta = 70.3°$. The above similarity relationships give

$$\frac{P/\pi a^2}{\sigma_0} = \alpha\left(\frac{\dot{h}/h}{\dot{\varepsilon}_0}\right)^{1/m}, \quad \alpha = \frac{F_a(m)}{\left[c(m)\tan\beta\right]^{1/m}}, \tag{4}$$

where the dimensionless parameter α is a monotonically increasing function of m. When $m=1$, the material is a Newtonian viscous solid and $\alpha = 4/3\tan\beta$ [6]. When $m \to \infty$, the material approaches the rate-independent limit, so that $\alpha = P/\pi a^2 \sigma_0$ is the constraint factor, being about 3 as shown in Fig. 1(a). For rate-independent solids, the true hardness values measured in nanoindentation tests by the Oliver-Pharr approach may be corrected for pile-up and sink-in effects using continuous stiffness measurement techniques and the known elastic modulus. For a rate-independent solid, we define $H_{\text{nominal}} = P/\pi(h\tan\beta)^2$ from the nominal projected contact area, $\pi(h\tan\beta)^2$. Unlike the true hardness, this quantity is experimentally convenient because of its simple relation to the depth of penetration. Consequently,

$$\frac{H_{\text{nominal}}}{\sigma_0} = \alpha^* \left(\frac{\dot{h}}{h \dot{\varepsilon}_0} \right)^{1/m}, \quad \dot{\varepsilon}_{\text{eff}} = \frac{\dot{h}}{h} = B H_{\text{nominal}}^m = A \left(\frac{H_{\text{nominal}}}{\alpha^*} \right)^m, \qquad (5)$$

where $\alpha^* = c^2 \alpha$. Comparing Eqs. (1) and (5) gives a ratio of $A/B = \alpha^{*m}$, which would allow one to determine the uniaxial creep parameter A from an indentation measurement of B.

Results in Fig. 1(a) are calculated using the available simulation data in Bower et al. [6], which are valid only for blunt indenters (flat-ended punches, or spheres in the limits of small displacement) as validated by our finite element results. For the Berkovich indenter,

$$H_{\text{nominal}} = \alpha_{\Delta}^* A^{-1/m} \left(\frac{\dot{h}}{h} \right)^{1/m}, \quad \frac{A}{B} = \alpha_{\Delta}^{*m}, \qquad (6)$$

where α_{Δ}^* can be determined from detailed finite element simulations. Only $m=1,2,5$ are simulated here. Comparisons in Fig. 1(b) show a negligible contribution of the contact shape to the A/B ratio, which justifies the use of the axisymmetric contact in theoretical studies.

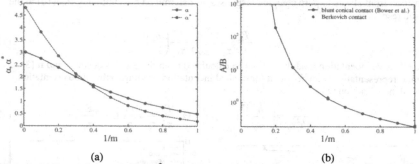

(a)   (b)

Figure 1. (a) Dependence of α and α^* on $1/m$ for Berkovich-equivalent conical contact [6]. (b) The A/B ratio for Berkovich-equivalent conical contact [6] and for Berkovich contact by finite element simulations.

INDENTATION OF CREEPING SINGLE CRYSTALS

Indentation responses of creeping single crystals can be represented in a form similar to Eq. (2). The classic version of crystal plasticity theory in [12,13] is followed in this work. Under infinitesimal deformation conditions, the plastic strain rates are assumed to be a sum of the slip rates over all the slip systems, namely,

$$\dot{\varepsilon}^p = \frac{1}{2} \sum_\alpha \dot{\gamma}^{(\alpha)} \left(s_i^{(\alpha)} m_j^{(\alpha)} + s_j^{(\alpha)} m_i^{(\alpha)} \right), \qquad (7)$$

where $\mathbf{s}^{(\alpha)}$ and $\mathbf{m}^{(\alpha)}$ are the slip direction and slip normal on the α-th slip system. The slip rate $\dot{\gamma}^{(\alpha)}$ relates to the resolved shear stress $\tau^{(\alpha)}$ by a power law,

$$\dot{\gamma}^{(\alpha)} = \dot{\gamma}_0 \left| \frac{\tau^{(\alpha)}}{\tau_0} \right|^m \text{sgn}\left(\tau^{(\alpha)} \right), \qquad (8)$$

211

where τ_0 is the slip strength, and $\dot{\gamma}_0$ is a reference strain rate. We assume no strain hardening, so that τ_0 is a constant.

For spherical indentation, choosing an effective strain rate, $\dot{\varepsilon}_{eff} = \dot{h}/a$, gives

$$\frac{P}{\pi a^2 \tau_0} = \left(\frac{\dot{h}}{a\dot{\gamma}_0}\right)^{1/m} F_a(m,\mathbf{n}), \quad h = \frac{1}{D}\left[\frac{a}{c(m,\mathbf{n})}\right]^2. \qquad (9)$$

The constants c and F_a are functions of the material constant m and the crystallographic orientation $\mathbf{n}$ of the half-space single crystal, and they depend weakly on the interface frictional condition. Eq. (9) can be rewritten as

$$\frac{P}{\pi D^2 \tau_0} = \left(\frac{\dot{h}}{D\dot{\gamma}_0}\right)^{\frac{1}{m}} \left(\frac{h}{D}\right)^{1-\frac{1}{2m}} \Theta, \qquad (10)$$

with $\Theta(m,\mathbf{n}) = F_a c^{2-\frac{1}{m}}$. The load-displacement curves from finite element simulations have a transition from elastic to pure creeping responses, as determined by comparing Eq. (10) to the elastic response:

$$\frac{P_{elastic}}{\pi D^2 \tau_0} = \frac{4}{3\sqrt{2}} \frac{E^*}{\pi \tau_0} \left(\frac{h}{D}\right)^{\frac{3}{2}}. \qquad (11)$$

The effective indentation modulus E^* can be calculated from the elastic constants and $\mathbf{n}$ [3]. Similar representations can be found for conical indentation, leading to the same orientation dependence function Θ in

$$\frac{P}{\pi h^2 \tau_0} = \left(\frac{\dot{h}}{h\dot{\varepsilon}_0}\right)^{\frac{1}{m}} (\tan \beta)^{2-\frac{1}{m}} \Theta, \quad \frac{P_{elastic}}{\pi h^2 \tau_0} = \frac{4}{\pi^2} \tan \beta \frac{E^*}{\tau_0}. \qquad (12)$$

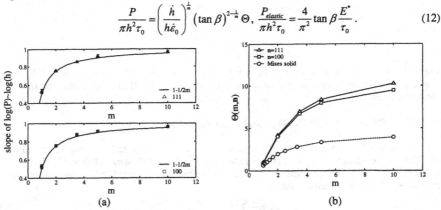

Figure 2. For spherical indentation on single crystals with $\dot{h} = const$, the slope of $\log(P) \sim \log(h)$ and the function Θ are plotted against the material constant m in (a) and (b), respectively. Two representative surface orientations are shown, $\mathbf{n} = \{111\}$ and $\{100\}$.

The above predictions have been validated by finite element simulations. The ABAQUS User-defined Material subroutine in [12] was modified according to the constitutive law specified in Eqs. (7) and (8). For a spherical indentation at a constant $\dot{h}$, the slope of a log-log plot of the load-displacement curve is predicted as $1 - 1/2m$, as confirmed by numerical results in Fig. 2(a). Results in Fig. 2(b) suggest that a plateau value is reached when m is about 10. Fig. 3 shows a variation of Θ of about 7% according to the contours plotted on the standard [001] triangle. The dependence on the indentation direction is thus weak for fcc single crystals because of the existence of many slip systems. Recent nanoindentation tests on coarse-grained aluminum specimens [14] seem to support this conclusion, since no noticeable variations except for experimental errors have been found when indenting various grains.

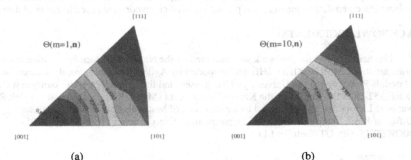

(a)   (b)

Figure 3. Contours of Θ, as plotted on the [001] inverse pole figure, show the dependence on surface normal for spherical indentation with constant $\dot{h}$: (a) $m=1$, (b) $m=10$.

EXPERIMENTAL COMPARISONS

The predicted A/B ratio has been compared to a number of experimental results. For amorphous selenium tested above the glass transition temperature (displaying a Newtonian viscous flow behavior and $m \approx 1$), the measured value is found to be $A/B \approx 0.256$ [7,9], while the predicted value is about 0.21 as shown in Fig. 1. For high-purity indium, Lucas and Oliver [8] found that $m \approx 7$ and $A/B > 10^5$. It should be noted that A/B goes to infinity as m increases from unity to infinity, as shown in Fig. 1. When using $\alpha^* = (A/B)^{1/m}$, we found that the experimental result gives $1.4 < \alpha^* < 1.7$, which is much less than the predicted α^* (about 3 in Fig. 1). A similar discrepancy is also noted in a recent work on nanocrystalline nickel [15], giving experimentally $m \approx 5$ and $\alpha^* \approx 1.4$, which again is much less than the predicted α^* (about 2.8 in Fig. 1). A close examination of these experiments suggests that this discrepancy might be due to the transient creep behavior. In the range of m values in [8,14,15], a pile-up is predicted while experimental observations show sink-in, which is a result of strain hardening behavior. The pure-creeping constitutive law in Eq. (1) is unable to describe this behavior, and more advanced analyses (e.g., [16]) are needed along this line.

SUMMARY

In summary, the correlations of contact pressure, effective strain, and effective strain rate in the indentation responses of creeping solids have been analyzed theoretically by a similarity approach and numerically by finite element simulations. The stress exponent (reciprocal to the strain rate sensitivity exponent) can be accurately determined, but other parameters, as indicated by the A/B ratio in Eqs. (1), (5), and (6), only agree with theoretical predictions when $m \rightarrow 1$. This discrepancy is likely to result from the strain hardening behavior in primary creep, while the theoretical predictions are limited to steady-state creep. These details may be important in using indentation creep data to measure and predict uniaxial creep constants used in material design.

ACKNOWLEDGEMENTS

Financial support for this work was provided by the National Science Foundation under grant number CMMI 0800168. JHL was supported by Agilent Technologies, the Center for Materials Processing at the University of Tennessee, and the Korea Research Foundation Grant (KRF-352-D00001) funded by the Korean Government (MOEHRD). Research at the Oak Ridge National Laboratory was sponsored by the Division of Materials Sciences and Engineering, Office of Basic Energy Sciences, US Department of Energy, under contract DE-AC05-00OR22725 with UT-Battelle, LLC.

REFERENCES

1. W.C. Oliver and G.M. Pharr, *J. Mater. Res.* **7**, 1564 (1992).
2. W.C. Oliver and G.M. Pharr, *J. Mater. Res.* **19**, 3 (2004).
3. Y.F. Gao and G.M. Pharr, *Scripta Mater.* **57**, 13 (2007).
4. Y.F. Gao, H.T. Xu, W.C. Oliver, and G.M. Pharr, *J. Mech. Phys. Solids* **56**, 402 (2008).
5. M.J. Mayo and W.D. Nix, *Acta Metall.* **36**, 2183 (1988).
6. A.F. Bower, N.A. Fleck, A. Needleman, and N. Ogbonna, *Proc R Soc London A* **441**, 97 (1993).
7. W.H. Poisl, W.C. Oliver, and B.D. Fabes, *J. Mater. Res.* **10**, 2024 (1995).
8. B.N. Lucas and W.C. Oliver, *Metall. Mater. Trans. A* **30**, 601 (1999).
9. J.A. LaManna, PhD Thesis, University of Tennessee (2006).
10. S.J. Sohn, PhD Thesis, University of Tennessee (2007).
11. F.X. Liu, Y.F. Gao, and P.K. Liaw, *Metall. Mater. Trans. A* **39**, 1862 (2008).
12. Y. Huang, Mech. Rep. 178, Division of Applied Science, Harvard University (1991).
13. J.W. Kysar, *J. Mech. Phys. Solids* **49**, 1099-1128 (2001).
14. C.J. Su, et al., unpublished experimental results (2009).
15. C. L. Wang, Y.H. Lai, J.C. Huang, and T.G. Nieh, *Scripta Mater,* **62**, 175 (2010).
16. N. Ogbonna, N.A. Fleck, A.C.F. Cocks, *Int. J. Mech. Sci.*, **37**, 1179 (1995).

Mater. Res. Soc. Symp. Proc. Vol. 1224 © 2010 Materials Research Society   1224-GG05-12

Effect of Surface Roughness on Determination of Creep Parameters Using Impression Creep Technique

Wuzhu Yan, B. Zhao and Zhufeng Yue
Department of Engineering Mechanics, Northwestern Polytechnical University, Xi'an, 710129, PR China

ABSTRACT

The indentation creep test, especially the impression creep, exhibits a magic appealing in the determination of creep properties of small structures in industry for its simplicity, efficiency and non-destruction merits. Most of previous researches of indentation or impression creep neglect the effect of surface roughness of materials, which plays a crucial role in extracting creep properties of materials. The FE results showed that the surface roughness has no effect on the determination of creep exponent when the punching stress is larger than 150MPa. However, under a smaller punching stress the stress exponent is decreased due to the "Tuner" effect of asperities. The conclusions drawn in the present study provide an important guidance on experiment results amendment for impression creep technique.

INTRODUCTION

. One of the crucial aims of the impression creep test is to bridge it and uni-axial tensile/compression creep test in the determination of creep parameters of materials [1-2]. The impression creep law of steady stage which is analogous with Norton law in format was exploited by Yu and Li [3] and is

$$\dot{h} = C'\sigma_m^{n'} \tag{1}$$

where $\dot{h}$ is the steady state punch velocity, n' is the stress exponent of impression creep, σ_m is the applied punching stress, and C' is the temperature and material dependent pre-exponential factor. However, most of investigations on impression creep including tests and numerical simulations neglected influence of surface roughness.

In the present study, the impression creep performed on a solid with a rough surface is investigated by numerical simulations to explore how the surface roughness works on determination of creep parameters.

FINITE ELEMENT MODEL

A solid pressed by a flat cylindrical punch with radius r can be simplified into a two-dimensional axial symmetric for the finite element calculation of impression creep. A uniform punching stress σ_m is acted on the tested material via the punch. The cosine surface profile is used to exhibits fractal behavior of surface in the simulation. In this paper, the surface roughness was quantified by:

$$R_a = \frac{1}{l}\int_0^l |y(x)| dx \qquad (2)$$

Where $y(x)$ is the height of the surface profile based on the centerline, l is the concerned length of the sample. Keeping the wave length fixed, the amplitude of the cosine curve was varied to generate R_a=0, 6.4, 12.7 and 25.5 μm respectively [4]. The finite element model with partially refined (at peaks and valleys) in the zone beneath the punch is designed as shown in Fig. 1.

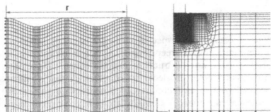

Fig. 1. Schematic of cosine profile surface and mesh design used in the numerical simulation of impression creep.

The typical structure steel mechanical properties is selected as a representation with Young's Modulus E=138GPa, Poisson's ratio μ=0.3. The Katchanov-Robotnov (K-R) model is employed to study the impression creep damage. Yue [5] has initially studied impression creep damage by implementing the K-R creep damage law into FE model, the creep damage Equations are:

$$\dot{\varepsilon}_e = \frac{B\sigma_e^n}{(1-\omega)^n} \qquad (3)$$

$$\dot{\omega} = \frac{D\left[\alpha\sigma_1 + (1-\alpha)\sigma_e\right]^X}{(1-\omega)^\phi} \qquad (4)$$

where ε_e and σ_e are the equivalent creep strain and stress, respectively. σ_1 is the maximum principal stress, and ω is the damage variable. The coefficients B, D, n, X, α and ϕ are material creep damage parameters. The parameter α controls different damage mechanisms (i.e. the relative importance between σ_1 and σ_e). Here we assume $\alpha = 0$, which is an acceptable simplification for most materials.

The K-R creep damage law is implemented into the finite element code ABAQUS as a user subroutine 'CREEP'. The creep damage parameters used in FE simulation are listed in Table 1 while for calculations without damage the creep parameters are $B = 1.7828 \times 10^{-17}$ s^{-1}/MPa^n, n=5.

Table 1. Creep damage model parameters for Eqs. 3 and 4.

$B\ [\text{s}^{-1}/\text{MPa}^n]$	1.7828×10^{-17}
$D\ [\text{s}^{-1}/\text{MPa}^X]$	1.0×10^{-17}
X	5.0
α	0
ϕ	5.0
n	5.0

RESULTS AND ANALYSIS

Effect of surface roughness on damage distribution

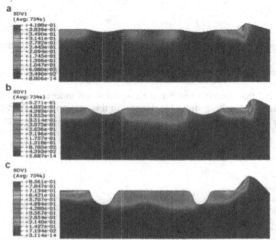

Fig. 2. Damage distributions for different surface roughness. (a) R_a=6.4 μm, (b) R_a=12.7 μm, (c) R_a=25.5 μm. σ_m=150 MPa, t=2×10⁵ s.

Fig. 2 shows the creep damage distribution for three different surface roughness, they are R_a=6.4 (Fig. 2a), 12.7 (Fig. 2b) and 25.5 μm (Fig. 2c), respectively. The other conditions (e.g. magnitude of punching stress and creep time) are the same for the three cases. From Fig. 2, three important conclusions can be drawn. Firstly, two obvious damage zones can be found. One is at the edge of the punch, the other is at peaks beneath the punch. Secondly, it is seen that a larger surface roughness R_a induces a higher damage level. The maximum damage ω_{max} for the three different surface roughness aforementioned are 0.42, 0.53 and 0.86 respectively, which occurs at the edge of the punch. This is further illustrated in Fig. 3, which shows the damage evolution at the edge of the punch for the three different R_a values. From Fig. 3, it can be seen that a higher R_a value induces a faster increase of damage. Moreover, a transient damage stage followed by a steady state can be observed throughout the damage evolution. Thirdly, the contact area A_c between the punch and the specimen decreased with the increase of surface roughness, which induces a larger net section stress* σ_N.

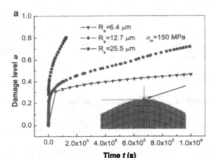

Fig. 3. Effect of surface roughness on evolution of creep damage.

Effect of surface roughness on penetration depth and punch velocity

It is reported that the effect of surface roughness on indentation behavior decreases with the increase of punching stress [6]. Hence, in this section, the effect of R_a values on punch velocity is investigated under a relatively smaller punching stress $\sigma_m=60$ MPa.

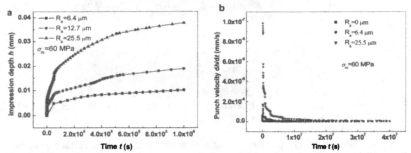

Fig. 4. Effect of surface roughness on evolutions of (a) penetration depth and (b) punch velocity.

Fig. 4a gives the evolution of impression depth h as a function of creep time t. It shows how the impression creep behaviors are influenced by surface roughness. It can be seen that the penetration depth increases with the increase of R_a. Under a smaller punching stress, we can not recognize the primary and the secondary creep stage all

---

* The net section stress used here indicates the net contact stress: $\sigma_N = \dfrac{\sigma_m \times \pi r^2}{A_c}$

218

through the impression creep process. The penetration depth *versus* creep time curves are consisted by a series of line segments whose slopes decrease with creep time. The reason is that the punching stress is so small that it cannot press the rough surface to be flat. As the damage accumulates, the net contact area increases, causing the impression net section stress σ_N to decrease. In addition, the decrease of impression depth rate[*] can be concluded.

Fig. 4b shows how the punch velocities associated with different surface roughness converge at a constant value with increasing of creep time. We can predict that the punch and specimen tend to full contact as $t \to \infty$, and the punch velocity will tend to a steady state, which is similar to that of perfectly smooth surface. In experiment, we generally assume the slope of the last segment of line as the steady state punch velocity, causing a larger measured steady state punch velocity over the theoretical one. This effect is apparent at a small punching stress.

Effect of surface roughness on determination of stress exponent of impression creep

From Eq. 1 we know that the slope of the $\lg \dot{h} - \lg \sigma_m$ curve is the stress exponent of impression creep. For the FE results, we apply the same numerical process to investigate the effect of surface roughness on impression creep exponent. In order to make an easier comparison, we consider R_a=0, 6.4 and 25.5 μm. For each case, the responses under the punching stress σ_m=30, 60, 100, 150, 200 and 250 MPa are investigated. The punch velocity at the end of the calculation is extracted from FE results. In order to observe the evolution of the impression creep exponent, $\lg \dot{h} - \lg \sigma_m$ curves are plotted for different R_a in Fig. 5.

Fig. 5 shows that the impression creep exponent is approximately equal to the one introduced into the FE model when the punching stress is larger than 150 MPa. It indicates that the effect of surface roughness on impression creep exponent is weak enough to be neglected under a larger punching stress due to the reach of flat surface and a steady state impression creep stage.

Under a smaller punching stress (<150 MPa), on the other hand, the derivate stress exponent is much smaller than the one introduced into the FE model. The phenomena can be rationalized as follows. Under a small punching stress, the net contact area between punch and specimen keeps expanding with the increase of damage. In this process, the rough surface plays a role of "Tuner" on the net section stress. For a relatively larger punching stress (note that it is still within the range of σ_m<150 MPa in the present study), a higher damage level can be induced. Hence, the net contact area increases sharply, inducing a sharp decrease of net section stress. For a relatively smaller punching stress, on the contrary, a lower damage level can be observed. Thus the net contact area increases slowly, causing a slow decrease of net section stress. As a result, the net section stress cannot change sharply under different punching stress due to their interaction. Additionally within the range of small punching stress (in the present study σ_m<150 MPa), the punch velocity is not very sensitive to the change of punching stress due to the "Tuner" effect of surface roughness, resulting in a smaller stress exponent of impression

---

[*] The impression creep rate used here is the one at the end of the calculation

creep. The larger the value of R_a is, the stronger the "Tuner" effect is, and the smaller the stress exponent will be until this effect vanishes.

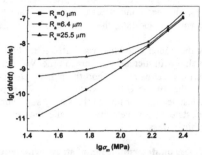

Fig. 5. Effect of surface roughness on determination of creep stress exponent.

CONCLUSIONS

The effect of surface roughness was investigated on determination of the stress exponent of impression creep while the K-R creep damage model was incorporated. The finite element results showed that the surface roughness caused a larger punch velocity than that of smooth surface for the same creep time, and a small punching stress could accelerate this trend. The larger the value of R_a is, the stronger the effect will be. The surface roughness would not affect the determination of the stress exponent of impression creep under a larger punching stress, while it can decrease the stress exponent a lot under a smaller punching stress.

ACKNOWLEDGEMENT

The authors appreciate the financial supports from National Nature Science Foundation of China (50775183, 50805118), Research Fund for Doctoral Program of Higher Education (N6CJ0001) and High-tech Research and Development Program of China (2009AA04Z418).

REFERENCES

[1] D.H. Sastry, *Mat. Sci. Eng.* **A** 409 67-75 (2005).
[2] B. Evans, *J Geophys Res.* **89 (10)** 4213-4222 (1984).
[3] H.Y. Yu and J.C.M. Li, *J. Mater. Sci.* 12 2214-2222 (1977).
[4] O. Hatamleh, J. Smith, D. Cohen and R. Bradley, *Applied Surface Science* **255** 7414 – 7426 (2009).
[5] Z.F. Yue, *Chinese Journal of Acta Metallurgica Sinica* **41** 15-18 (2005).
[6] W.Z. Yan, S.F. Wen, J. Liu and Z.F. Yue, *Acta Metall. Sin. (Engl. Lett.)* **22 (6)** 468-480 (2009).

Mater. Res. Soc. Symp. Proc. Vol. 1224 © 2010 Materials Research Society   1224-GG04-07

Effect of dilatation on the elasto-plastic response of bulk metallic glasses under indentation

Anamika Prasad[1], Ming Dao[2], and Upadrasta Ramamurty[3]

[1]Department of Bioengineering, Stanford University, Palo Alto, CA, USA
[2]Department of Materials Science and Engineering, MIT, Cambridge, MA, USA
[3]Department of Materials Engineering, Indian Institute of Science, Bangalore, India

ABSTRACT

Unlike metals, elasto-plastic response of bulk metallic glasses (BMGs) follows closely that of granular materials through pressure dependent (or normal stress) yield locus and shear stress induced material dilatation. While on a micro-structural level, material dilatation is responsible for stress-softening and formation of localized shear band, its influence on the macro-scale flow and deformation is largely unknown. In this work, we systematically analyze the effect of material dilatation on the gross indentation response of Zr-based BMG via finite element simulation. The strengthening/softening effect of load-depth response and corresponding stress-strain profiles are presented in light of differences in elastic-plastic regimes under common indenters. Through comparison of the numerical predictions with existing experimental data, we draw conclusions to guide the selection of a suitable dilatation parameter for accurately predicting the gross response of BMGs.

INTRODUCTION

Bulk metallic glasses (BMGs) have gained considerable scientific and practical importance due to their unique combinations of mechanical properties [1]. This has opened up new avenues for their structural applications [2] and hence greater need for understanding the mechanical response of BMGs at different length scales, and for developing accurate predictive capabilities. On a macro-scale level, it is now commonly agreed that their plastic yield condition is better predicted using the pressure sensitive (or normal stress sensitive) *yield criteria* such as the Drucker-Prager (or the Mohr-Coulomb) model [3]. However, the *flow rule* or the plastic flow condition beyond yield is not clearly agreed upon and is the focus of the current study. We systematically investigate the effect of different plastic flow conditions for the Drucker-Prager (DP) yield criteria. Specifically, we focus on the effect of variation in dilatation angle on the gross indentation response of BMGs through numerical simulation and compare the predictions with existing experimental data to guide the selection of a suitable material parameter.

FINITE ELEMENT AND THE MATERIAL MODEL

Conical (apex angle of 90^0 and 140.6^0) and pyramidal indentation (Berkovich and Vickers indenter) analysis was performed using the general-purpose FEM package ABAQUS Standard (SIMULIA, Providence, RI, USA). Axisymmetric two-dimensional mesh was used for conical indentation, while a full three-dimensional mesh was used for pyramidal indentation, with the mesh size being optimized for result accuracy. For all cases, at least 10 to 15 elements were in contact with the indenter at full indentation depth. A typical mesh for the axisymmetric case is shown in Figure 1a.

The extended DP material model of ABAQUS was used for BMGs, where the *yield function (f)* and the *flow rule (g)* is given by equations (1) and (2) below and is shown in Figure 1b. The p-q space is related to the stress invariants space, where p relates to the first invariant of principal stress, and q relates to the second invariant of deviatoric shear stress. The material parameters include the material friction angle (β), the cohesive strength (d), and the dilatation angle (ψ).

$$f(p,q) = q - p\tan(\beta) - d = 0 \qquad (1)$$

$$g = q - p\tan(\psi) \qquad (2)$$

The dilatation angle determines the inclination of plastic flow from the shear stress direction q (see Figure 1b). The *associative* (or the *normality) flow rule* assumes a flow direction normal to the yield function, as indicated by the dashed arrow in the figure, whereby ψ=β. *Non-associative flow rule* on the other hand allows for an independent representation of the plastic flow and is indicated by the solid arrow in the figure. Here, material parameters for Zr-based BMG were taken from literature [3] corresponding to non-associated DP model, with the dilatation angle being varied as percentage of β from 0% to 100%, where 100% corresponds to associated flow condition.

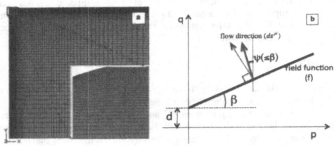

Figure 1: (a) Typical finite element mesh for the axisymmetric indentation with the inset showing refined mesh close to indentation (b) Drucker-Prager yield function in p-q space. The material parameters β and d represent the material friction angle and cohesion, respectively. The dilatation angle is given by ψ, where ψ=β represents the normality or the associative flow condition.

RESULTS AND DISCUSSION

The results are discussed in light of typical plastic flow fields below common indenters, as shown in Figure 2, where the equivalent plastic strains (PEEQ) for the sharp conical (90°) and the Berkovich indenters is shown under similar indentation load of 10 N. The plastic field demarcation zones are based on work by Giannakopolos and Suresh [4], where PEEQ> 29% denotes the *cutting zone* where intense rigid plasticity flow is expected. For the sharp conical indenter, the cutting zone completely surrounds the indenter and reaches the free surface outside the indentation ring, while for the Berkovich indenter, this zone lies close to the indentation tip. The overall plasticity zone is confined to a much smaller zone for the 90° cone (< 2h, where h is the indentation depth) as compared to Berkovich case (>5h). The profile for the other indenters considered here (conical 140.6° and vickers) follow similar trends as for the Berkovich case. The load-depth and material pile-up response will be affected by the plastic zones of indentation and hence will guide the interpretation of the results that follow.

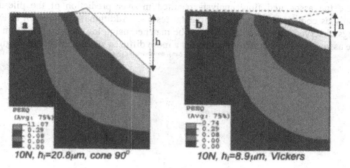

Figure 2: Typical plastic zones under indentation for pressure sensitive material with no dilatation with (a) sharp conical indentation of 90°, and (b) Berkovich indenter. The zones are demarcated based on Giannakopoulos and Suresh [4] where PEEQ>29% indicates cutting zones with rigid plastic response. The response of the sharp conical indenter is distinctly different from others in that rigid-plasticity completely surrounds the indenter and reaches the surface of indentation.

The indentation load-depth profile provides valuable information for the mechanical response of the surface and is commonly used to extract the elasto-plastic material parameters. Hence it is important to study the influence of dilatation parameter on the P-h curve and is shown in Figure 3a, where the results are plotted for the sharp conical and the Berkovich indenter with no (0%), low (10%) and full (100%) dilatation. As can been seen in the figure, there is a stiffening of the load-depth response with increase in dilatation for both the indenters, with the results showing low sensitivity to dilatation angle ≤ 10%. Overall, the normalized load increases by 10% to 20% for the associated flow condition compared with corresponding case with no dilatation, the higher value being observed in the 90° cone. The pile-up response shows similar trend of increasing value with increasing dilatation, with the pile-up being higher by a factor of 2 for 90° conical indenter.

These effects and the differences in behavior between the indenters can be explained via the localized compressive nature of indentation and through the differences in plastic flow fields below different indenters. The increase in the dilatation angle causes an increase in the amount of material flow, which in turn is resisted by the relatively stiffer elastic core outside the plasticity zone and hence leading to the stiffening of the load-depth response. The pile-up response on the other hand is affected by the nature of plastic flow closer to the indenter. Since the cutting zone completely surrounds the sharp conical indenter, it shows a much higher pile-up and also increased sensitivity of the pile-up value to the material dilatation angle.

The above differences in the load-depth and pile-up response with dilatation angle can have important implications for property extraction and hardness prediction of BMGs. Based on direct density measurements of BMGs, the overall volume reduction for the material is typically low and is of the order of 0.15-0.25% [1]. Hence BMGs show low gross dilatation effects unlike other granular materials. Thus with associative flow rule assumed in numerical predictions, the fitting of the load-depth curve to experimental response may lead to gross error in property estimation and/or over prediction of the material pile-up. This likely discrepancy in experimental and numerical results with associated flow analysis is supported by the work by Kervyin et al [5], where the associated flow analysis resulted in over prediction of the pile-up response numerically as compared to their experimental results. Additionally, as shown in Figure 4, the shear rings observed experimentally [3] can be matched using low or zero dilatation (figure 4a), while the associative flow conditions predicts a more diffused surface effect (see Figure 4b).

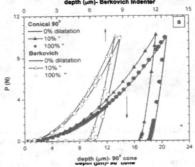

Figure 3: (a) Load depth response with and without associative flow rule for sharp conical and Berkovich indenters. Clearly, associative flow rule has significant stiffening effect on the P-h curve with the influence being more prominent for sharp conical indenter.

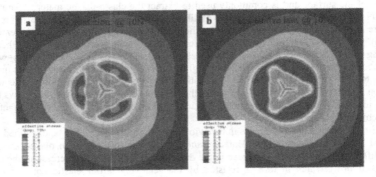

Figure 4: Effective Von mises stress on the surface of the material for Berkovich indenter for material with (a) zero dilatation, and (b) associative flow rule. The associative flow rule predicts a diffused surface plasticity, while with zero dilatation, the results follows closely the shear bands rings observed experimentally.

SUMMARY AND CONCLUSION

Using finite element analysis of indentation, the influence of dilatation parameter on the numerical prediction of pressure-sensitive BMGs is investigated and the results are discussed in light of experimental data for Zr-based BMGs. We have shown that careful attention to the flow rule is required to answer some of the discrepancies currently present in literature in terms of matching the experimental results to numerical prediction. Overall, a low or zero dilatation is preferred, given the low level of dilatation in experiments. Associated flow condition is clearly not accurate for BMGs as it can lead to as high as 20% stiffer load-depth response and increase in pile-up by a factor of 2. It is also important to note here while simple models such as DP can incorporate gross dilatation effects, sophisticated material models such as the elastic-viscoplastic model of Anand and Su [7] can more accurately predict the frictional and dilatant mechanism of plastic flow of BMGs. However, in spite of these limitations, these simple models will continue to play an important role for BMG modeling given their ability to capture the gross mechanical response, ease of use, and availability in most commercial FE codes.

Finally, we should also note that continuum mechanics based FEM simulations including the sophisticated constitutive models do not include atomistic level details and related smaller scale mechanics. For understanding detailed atomistic level mechanisms, finer scale modeling/simulations should be pursued.

REFERENCES

1. C.A. Schuh, T.C Hufnagel, and U. Ramamurty, Acta Mat. **55** (12), 4067-4109 (2007).
2. M. Ashby and A. Greer, Scripta Mat. **54** (3), 321-326 (2006).
3. R. Vaidyanathan, M. Dao, G. Ravichandran, S. Suresh, Acta Mat. **49** (18), 3781-3789 (2001)
4. A.E Giannakoulos and S. Suresh, Scripta Mat. **40** (10), 1191-1198 (1999).
5. V. Keryvin, R. Crosnier1, R. Laniel, V.H. Hoang, and J.C. Sangleboeuf. J. Phys. D: Applied Phys. 41, (2008).
6. A. Prasad, R. Raghavan, S. Bellemare, M. Dao, S. Suresh, and U. Ramamurty (under progress journal article).
7. L. Anand and C. Su, Journal of the Mechanics and Physics of Solids, 53, 1362–1396 (2005).

Mater. Res. Soc. Symp. Proc. Vol. 1224 © 2010 Materials Research Society   1224-GG05-22

Stumbling on Extrinsic Effects in Super-hard Nanobuttons

Antonio Rinaldi*[1,2,3], Pedro Peralta[3], Cody Friesen[3], Dhiraj Nahar[3], Silvia Licoccia[1], Enrico Traversa[1,4], and Karl Sieradzki[3]

[1]NAST Center & Department of Chemical Science and Technology; Universita' di Roma "Tor Vergata", Via della Ricerca Scientifica, Roma, 00133, Italy (*antonio.rinaldi@uniroma2.it).
[2] ISPESL, Dipartimento Tecnologie di Sicurezza, ISPESL Via Alessandria 220/e, 00198, Italy.
[3] Fulton Schools of Engineering, School of Mechanical, Aerospace, Chemical and Materials Engineering, Arizona State University, Tempe, AZ 85287-6106, USA.
[4]International Research Center for Materials Nanoarchitectonics (MANA), National Institute for Materials Science (NIMS), 1-1 Namiki, Tsukuba, 305-0044, Ibaraki, Japan.

ABSTRACT

The compressive plastic strength of nanosized single crystal metallic pillars is known to depend on the diameter D, but little attention has been given to the pillar height h. The important role of h is analyzed here, observing the suppression of generalized crystal plasticity below a critical value h_{CR} that can be estimated a priori. Novel in-situ compression tests on regular pillars (D = 300-900 nm) as well as nanobuttons (i.e. very short pillars with h less than h_{CR}, such as D = 200 nm and h < 120 nm in this case) show that the latter ones are exceedingly harder than ordinary Ni pillars, withstanding stresses greater than 2 GPa. This h-controlled transition in the plastic behaviour is accompanied by extrinsic plastic effects in the harder nanobuttons. Such effects normally arise as Saint-Venant's assumption ceases to be accurate. Some bias related to those effects is identified and removed from test data. Our results underline that nanoscale testing is challenging when current methodology and technology are pushed to the limit.

INTRODUCTION

Experiments on micro- and nanopillars have become of particular interest after innovative micro-compression experiments were designed and performed a few years back on single-crystal metal pillars micro-machined by focused ion beam (FIB) [1]. That allowed demonstrating the effect of the external length scale - i.e., the pillar diameter D - on crystal plasticity over an unprecedented "broad" size range. By varying D from tens of micrometers to a few hundred nanometers, it was possible to clearly highlight the much higher compressive plastic strength of nanometer-scale single crystal pillars compared to bulk values found in "conventionally" sized samples. The strain-gradient free configuration (neglecting the axial tapering of the pillars) in these tests allowed ruling out immediately the class of "gradient plasticity" theories as a possible basis for the observed size effects [2]. Several studies on pillars followed, which also examined other material systems such as polycrystalline Ni pillars [3]. A recent, rather comprehensive review about pillar experiments appeared recently [4].

Almost invariably, though, the focus has been on the characterization of the D-dependence of plastic and strength properties, while the role the pillar height "h" seems to have passed unnoticed, despite the fact h should not be regarded as a minor parameter at all. If θ is the angle between the slip plane normal vector and the pillar axis, crystal plasticity is expected to be constrained when h drops below a critical height "$h_{CR} = D \tan(\theta)$" below which no slip plane would fully extend over the entire cross section. In terms of the aspect ratio "$AR = h/D$", h_{CR} would correspond to the critical threshold value "$AR_{CR} = \tan(\theta)$". This expectation has never been confirmed experimentally before.

Here, new tests to assess the proposed h-effect are devised, implemented, and analyzed for the first time by means of a refined methodology [5] suitable for performing highly precise *in-situ* compression tests of ordinary pillars as well as short pillars with "$AR < AR_{CR}$" and $D \sim 200$ nm. Hereafter, the word "pillar" is reserved for nanostructures with "$AR > AR_{CR}$", whereas the term "buttons" is introduced to indicate shorter pillars. When lowering h below h_{CR}, we show that it is possible to effectively suppress generalized plasticity by "switching off" the typical deformation mechanism of crystal plasticity from a global standpoint. As a consequence, buttons are harder materials compared to pillars and extrinsic effects become clearly observable, as they are amplified and contribute to a large fraction of the overall measured strain during the compression test.

EXPERIMENTAL DETAILS

Pillars with diameters of order 450-900 nm were machined with the FIB (FEI series Nova 200) in the surface of a grain of a polycrystalline Ni sample that had been, annealed, mechanically polished, and re-annealed again to remove the dislocations and the residual plastic deformation before ion milling. The crystallographic orientation **[0.303, 0.185, 0.935]** of the surface normal of that grain was determined via Electron Backscattering Diffraction (EBSD) in the SEM. Buttons with $D \sim 200$ nm and $AR = 0.4$-0.6 were machined next in the same grain. Note that, because $\theta = 52°$, the threshold aspect ratio was expected to be $AR_{CR} = 1.6$. The FIB machining (at 20 kV) was done in several passes at current levels ranging from 100 pA (first rough pass) to 10 pA (for pillar profiling) to control the final shape and to limit the amount of ion-induced damage. SEM was used to characterize the geometry of the columns before and after the compression and to transform the force *vs.* displacement curves into engineering stress-strain curves. A search technique was developed to locate the target features with the nanoindenter, as reported elsewhere [5]. A Hysitron© Triboscope nanoindenter was used for mechanical testing, which was conducted in force controlled mode. The mechanical tests were performed with a special indenter tip created "in house" by profiling a standard cube-corner tip with the FIB in such a way to have a 2 microns flat plateau with a small excrescence on one corner that could act as a scanning "tooth". This geometry allows for both SPM imaging of a nano-feature by the corner tooth and compressing the same feature under the flat surface after performing a proper offset to move the scanning tooth away from the contact area [5]. Figure 1 displays both the nanobutton image of a nanobutton and the corresponding SEM image. Ancillary experiments were performed in order to determine the sensitivity of this instrument to slip events [3]. The root mean square (RMS) noise on the force-displacement signal was 0.15 nm in the displacement and 0.26 μN in the force (these values refer to the vertical motion of the indenter tip), which made it possible to collect "sub-nanometer scale" precise experimental data and detect subtler extrinsic effects. Raw stress-strain datasets were corrected for the Sneddon effect to remove the contribution of the elastic deformation of the compliant foundation from the measured strain [3,4]. In fact, the raw force-displacement curve would account for the deformation of both the nanotarget and the elastic foundation underneath (while the deformation of the stiff tip is discarded). The fraction of the deformation associated to the foundation amounts to a very large percentage for low *AR* and becomes greater than 50% for nanobuttons.

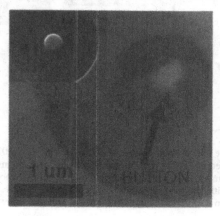

Figure 1. A button as imaged by the custom made tip *vs.* the SEM micrograph (in the inset).

RESULTS AND DISCUSSION

The h-controlled transition in crystal plasticity

Figure 2 displays representative stress-strain datasets (after correcting for the Sneddon effect). The insets show SEM images of a button (before testing) and of a 900 nm pillar (after testing). The plotted responses confirm at a glance that pillars and buttons have distinct behavior. Pillars experienced a plastic deformation starting at around 500 MPa with size-dependent strain hardening. Each pillar response is a series of horizontal strain bursts of crescent magnitude that culminated with a final indefinite plateau marking the failure and happening between 500 MPa and 1 GPa depending on *D*. Compressive failure in single crystal metals was accompanied by usual slip lines localization (see pillar in the inset). On the other hand, the button was much stronger. It sustained stresses > 2 GPa and behaved nearly linearly during loading without failure (not shown). The button only suffered a local deformation at the contact surface, associated to the tip-induced damage caused by our specially shaped diamond tip during SPM scanning, before and after testing [5].

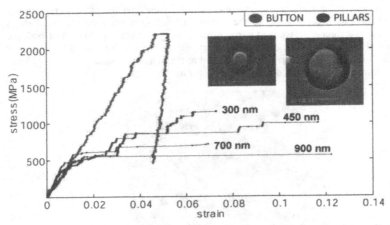

Figure 2. Representative stress-strain curves of a button with $D \sim 200$ nm and $AR \sim 0.6$ vs. several pillars with different D. The former exhibits a distinct response, nearly linear during loading up to the maximum load and without failure. The pillars show the well-known size dependent plasticity, with ultimate strengths ranging up to 1 GPa. The insets show SEM micrographs of the button (before test) and of one 900 nm pillar (after test).

The super-hard response of the buttons is owed to the equilibrating action of the base that acts as a kinematical constraint and prevents slip from happening. Even though the slip plane prolongs inside the foundation, the base of the button always acts as an effective barrier for dislocation motion due to two factors: *(i)* the much higher energy required by a dislocation to move in the bulk as opposed to that needed to egress to a free surface (*e.g.* due to a lack of image forces), and *(ii)* the rapid decay of compression and shear stresses in the foundation underneath the button and approaching a Boussinesq-type of distribution in the far field.

We speculate that failure in the nanobuttons may perhaps only happen if a different failure mode gets activated at stresses much higher than the pillar strength, unless the foundation fails first. For now, the actual compressive strength of the buttons remains unknown and could be tremendously higher, especially when considering that the limit value for compressive strength ($\sim G / \pi$) of a defect-free Ni crystal should theoretically exceed 20 GPa.

These data prove the existence of the h-controlled transition from plastic pillars to super-hard buttons, which enriches and completes the phenomenology of size-dependent plasticity with a mechanism that can be useful for MEMS/NEMS design purposes.

Discrete Extrinsic Effects

A closer examination of the button response in Figure 2 reveals the presence of an inelastic behavior, despite the constraint of crystal plasticity. In fact, the mismatch between the loading and the unloading paths is a clear indication of inelastic deformation. Such a deformation is instead associated with local and atomic plastic effects. As a proof, in Figure 3a we magnify the button data from Figure 2 and plot another dataset from a second nanobutton ($D \sim 190$ nm, $AR = 0.57$). The two stress responses display a striking agreement up to a stress level of about 1 GPa and then depart from each other due to a series of visible plastic slips of different magnitude. Note that the largest slip event observed in Figure 3a is too modest of a plastic displacement (*i.e.* less than 1 nm) to represent generalized crystal slip. Interestingly, these "bursts" were found not to occur at random but rather to happen orderly, in a systematic pairwise fashion and at defined stress ranges, in both buttons (paired slips on the two curves are pointed out in the inset of Figure 3a). This is a meaningful clue about the causal nexus underlying this deformation. Such small and ordered slips could likely originate from the local "engaging" between the tip and the top of the button. Reportedly, the roughness of the contact surfaces and the difference in hardness of diamond and nickel are likely dislocation sources [6]. Alternatively, the slips could relate to the plastic deformation of the compliant foundation under the pressure exerted by the harder button (recall that the applied Sneddon correction only removes the elastic component of the foundation strain from the experimental data, while the plastic component -if any- would still appear in Figure 3a). In either case, those plastic slips are not believed to represent any intrinsic button property, but are rather interpreted as extrinsic bias in the system response from plastic events happening at the button ends or externally. In support of this thesis, if the slips were removed from the curves, the remaining data would yield the two smooth curves shown in Figure 3b, closely matching together beyond 1 GPa all the way up to ~1.7 GPa. Our interpretation seems consistent from a theoretical standpoint when acknowledging that Saint-Venant's principle is not applicable with sufficient accuracy in this situation.

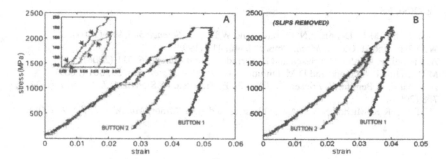

Figure 3. (A) Responses of two buttons of the same size (*i.e.* D and AR) loaded above 2 GPa (blue) and at around 1.7 GPa (red). Data are in close agreement during loading up to above 1 GPa. The button response is quasi-linear between two consecutive plateaus, with nearly identical slopes for both buttons – in fact the two datasets in the inset seem actually piecewise offset. (B) The pseudo-responses for the two buttons obtained by removing the detectable slip events from the original data now agree throughout the loading range.

CONCLUSIONS

As far as intrinsic effects, these tests rendered the first experimental proof of the existence of a transition from plastic pillars to super-hard buttons associated with the constraint of crystal plasticity below a critical value of height h_{CR}. Apart from intrinsic effects, the constraint of crystal plasticity offered the opportunity to emphasize the presence of inelastic extrinsic effects that usually gain importance when the Saint-Venant's assumption ceases to hold. From a metrological standpoint, the results suggest that testing nanostructures similarly sized or smaller than our nanobuttons is challenging given the present technological state, on one hand owing to the need for equipment with higher load resolution to discern atomic-scale plastic events, and on the other one due to extrinsic effects that need to be thoroughly understood and filtered out from raw experimental data when testing super-hard nanomaterials like nanobuttons or other nanomaterials (*e.g.* nanoparticles, CNTs, etc.). The debiasing strategy used to remove the inelastic slip events in nanobuttons - in addition to the usual Sneddon elastic correction - was a first move in that direction.

ACKNOWLEDGMENTS

The authors gratefully acknowledge that support of this work was provided by the Ira A. Fulton School of Engineering of Arizona State University. The access to the facilities in the LeRoy Eyring Center for Solid State Science at Arizona State University is also acknowledged.

REFERENCES

1. M.D. Uchic, M.D. Dennis, J.N. Florando and W.D. Nix, *Science* **305**, 986 (2004).
2. W.D. Nix, and H. Gao, *J. Mech. Phys. Sol.* **46**, 411 (1998).
3. A. Rinaldi, P. Peralta, C. Friesen and K. Sieradzki, *Acta Mater.* **56**, 511 (2008).
4. M.D. Uchic, P.A. Shade and D.M. Dimiduk, *Annual Rev. Mater. Res.* **39**, 361 (2009).
5. A. Rinaldi, P. Peralta, C. Friesen, N. Chawla, E. Traversa, K. Sieradzki, *J. Mater. Res.* **24**, 3, 768 (2009).
6. W. Shan, R.K. Mishra, S.A.S. Asif, O.Warren and A.M. Minor, *Nat. Mat.* **7**, 115-9 (2008).

AUTHOR INDEX

SUBJECT INDEX

Printed in the United States
by Bookmasters

Printed in the United States
By Bookmasters